Discovering the FULLNESS OF REALITY

How Partial Truths Obscure the Union of Faith and Science

David J. Keys, PhD

En Route Books and Media, LLC
Saint Louis, MO

ENROUTE
Make the time

En Route Books and Media, LLC
5705 Rhodes Avenue
St. Louis, MO 63109

Cover credit: Sebastian Mahfood

ISBN-13: 978-1-952464-85-0
Library of Congress Control Number: 2021940385

Table of Contents

Introduction		i
Chapter 1	Let's Get Real	1
Chapter 2	In the Beginning	21
Chapter 3	What a Start	43
Chapter 4	Following Through	55
Chapter 5	If He Builds it, They Will Come	75
Chapter 6	Time Won't Let Me	95
Chapter 7	You Say You Want an Evolution	107
Chapter 8	Just Who Do We Think We Are?	131
Chapter 9	What Kind of Science is Conscience?	167
Chapter 10	Sense or Nonsense?	177
Chapter 11	A Cloak for All Seasons	201
Chapter 12	How About a Little Team Spirit	225
Chapter 13	Removing the Blinders	253
Appendix A	Discussion Questions	277
Appendix B	List of Figures	307
Appendix C	Suggested Reading	311
Index		315

Introduction

I'm GUILTY. You're GUILTY. We all are, or at least have been, GUILTY at one time or another. We are all GUILTY of willfully failing to see truth because we choose not to. This doesn't mean we necessarily actively choose to override the truth, although some may do this—ever hear of a politician? More often, however, we have simply chosen, consciously or unconsciously, to not investigate the truth and may not even be aware that we are choosing to ignore truth. We choose to be willfully blind to the truth for a variety of reasons. In the worst cases, some people are willfully blind because they know that if they accepted the truth, they would have to change their lives—including perhaps, their jobs, their relationships, their habits, their choices.

However, I don't believe most of us are such minded. Instead, sometimes we are willfully blind to the truth because we think we already know the truth, or at least so we think we do. In many cases, we think we know the truth simply because we were raised as children to believe certain things. We trusted, and still do, our parents and others, who have taught us. Trust is a good thing, but perhaps when your parents/teachers learned their truths, they misunderstood, or the information was in error, or the information no longer applies. Perhaps there is better information or a better understanding now, or, perhaps, conversely, today's information is false, and the past truth is the real truth! Trust is an essential pre-requisite to knowledge as all knowledge comes to us because of Trust. We trust our senses, our teachers, our reason, those who love us, even the equipment that makes the measurements used by scientists and others to determine truths. But Trust does not make it true. So, what should we do?

This is quite a dilemma! As a human, we are wired to seek truth. Often people are willfully blind because of Truth. Unfortunately, it is Partial Truth that gets in the way. This must be overcome. Science tries to seek truth and thereby, come to a greater understanding of how things work. Sciences, such as physics, chemistry, biology, etc., deal specifically with the physical realities of life. The Arts, however, also seek truth and try to convey the ultimate expression of the non-physical realities of life such as beauty, love, emotions, etc. All, including both atheists and theists, in the end, seek the ultimate truths—"Who am I?" "What am I?" "What is the meaning of life?"

But here is the clinker. No one has the time, money, expertise, knowledge, etc., to fully pursue all of Truth. After all, we have a life to live, children and others to care for, jobs to do. We are limited. We can only learn part of the truth through our lives and through the lives and works of others. There is a lot to be learned! We willfully choose what is important to us and what is not. Let's get today done and move on to tomorrow.

Still, for everyone there will always be those moments, perhaps when things are not going as one would wish, in which one finally asks, "What's this all about?" Indeed, there may be times when something new comes along that doesn't fit into what we and others we trust have come to believe. When that happens, we need to re-evaluate the knowledge we hold on to.

So, I come to the reason for writing this book. I have found that many people want to know more but just can't find the energy or time to pursue truth, though the interest is there. I will explain what I know and hope that you may perhaps gain a deeper sense of Truth. But, what Truth? A zillion books couldn't hold all the truths in the world. The specific Truth which I intend to address is—"Is there a transcendent element to Human Life?" I don't think I am giving much away if I say upfront that I have come to know that there is a transcendent side of life, beyond all physicality. I hope to show you what I mean by that.

For that purpose, this book is intended to present to the reader both things of the world which exist and affirm the teachings of science, likewise, things of this world which apparently defy such teaching. Similarly, I will present both things of the world which affirm the teachings of religion and things of the world which apparently defy such teaching. As by definition, Truth cannot deny another truth. I hope to show that, in the end, apparent contradicting truths can both be true, but true in the sense they both accurately describe an aspect of reality. It is when one considers the FULLNESS OF

REALITY that apparent opposing truths can be recognized as addressing only a partial reality. With this new understanding, apparent contradictions can be reconciled to one another. So, in the end, I hope you find that science and faith, which many say do not mix, do not oppose one another after all; rather, they complement each other.

Chapter 1
Let's get real

There are those in society who by virtue of their intelligence, by their ability to have insight into matters which others simply fail to grasp, and by the respect shown to them by others in their profession are sought after for guidance in the meaning and intricacies of life. Albert Einstein was such a person. Einstein was a man who saw a reality in the world of physics that, certainly at first, few in his profession understood. He spoke of the inherent nature of space, time, and matter. Einstein's view of the world around us is far different from what the everyday person perceives. Things that we take for granted, like the constancy of time, length, and mass, he found not to be constant at all. They all depend on one's so-called frame of reference. What is true in one frame is not true in another. Reality in the material world is all relative.

Yet, in spite of this relativistic view, Einstein found the search for truth to have great meaning. He did not see Truth as being relative. Einstein saw the necessity of both Science and Religion. He once said, "Science can only be created by those who are thoroughly imbued with the aspiration toward truth and understanding." He continued, "This source of feeling, however, springs from the sphere of religion." He very much believed, "To this there also belongs the faith in the possibility that the regulations valid for the world of existence are rational, that is, comprehensible to reason. I cannot conceive of a genuine scientist without that profound faith. The situation may be expressed by an image: science without religion is lame, religion without science is blind."[1]

So, what did Einstein think of the reality in which we all find ourselves immersed? It has also been said by some that Albert Einstein once remarked, "*Reality is but an illusion, albeit a persistent one.*" The quote implies that what we think is real may not be at all. Albert Einstein also spoke of the illusion of TIME. He said, "*For us believing physicists, the distinction between past, present and future is only a stubbornly persistent illusion.*"[2] The "believing physicists" have nothing to do with a faith-based religious believing; instead, this deals with the physicist's view of what time really is.

---

[1] 'Science, Philosophy and Religion,' prepared for initial meeting of the Conference on Science, Philosophy and Religion in Their Relation to the Democratic Way of Life, at the Jewish Theological Seminary of America, New York City (9-11 Sep 1940). Collected in *Albert Einstein: In His Own Words* (New York: Random House, 2000), 212.

[2]Calaprice, Alice, ed. *The Expanded Quotable Einstein.* (Princeton: Princeton University Press, 2000), 75.

So how does one understand anything in life if all life is "but an illusion"? Let's look into this quandary a little bit. For physicists, time is not a concrete, fixed entity. What is one minute for you is not necessarily one minute in another time frame. However, the everyday person has no such issues. We know what time is, or at least we think we know. One year for me should be the same as one year for you. Seems pretty clear to me.

There is, however, a perplexing problem in physics called the Twin Paradox, which, in my version, goes something like this. Once upon a time in a galaxy far, far away (oops, wrong story). Once, there were two twins who were twenty years old. One twin, Tom, became an astronaut. His brother became an accountant. Tom left his pet dog Fido with his twin brother, Karl, and set out on an interstellar journey where he traveled at 87% of the speed of light. Twenty Earth years later, as determined by Tom's watch, the forty-year-old Tom returned to Earth and went to visit his brother Karl. He found that Fido, of course, was long gone, and his brother Karl was now grey haired and according to Karl's calendar was sixty years old. How can that be? Do you believe it could happen, or do you just think I am making that up? It seems absurd. After all, I wasn't born yesterday. Where is the punch line? But the fact is, it would be true. Experiments have shown that time does depend on your frame of reference. For Einstein, the past, present, and future of time need not be distinguishable. In some ways, it sounds almost conceptually more in tune with religion. If time is not absolute and is not the same for all of us, maybe it is not such a big deal for God to be outside of time. It seems our view of reality is limited to the very narrow conditions in which we live.

Here are two everyday examples[3]:

- GPS navigational systems must take into account the time dilation due to the different velocity of the satellites compared to a location of a particular spot on Earth. This difference in speed is due to the fact that the satellites are about 12,600 miles above the Earth. As the Earth surface rotates, the satellites must rotate even faster to keep up and stay over the same spot on Earth. This time dilation plus some gravitational effects amount to about 7,000 nanoseconds a day. The satellites themselves have built in clocks which are accurate to a few nanoseconds. Without correcting for these effects, the calculation of the location of the nearby gas station would be off by an additional five miles each day!
- Remember those old TVs you grew up with? The electron beams had to be fired exactly at the TV screen at the right pixel in order to show the correct

[3] Jesse Emspak, "8 Ways You Can See Einstein's Theory of Relativity in Real Life." atwww.livescience.com/58245-theory-of-relativity-in-real-life.html

> red, green, or blue color. The beam then had to be quickly moved to the next spot. The speeds of all this were sufficiently fast that the magnets guiding the direction of the beams had to be specifically designed to take into account the relativistic effects of the change in mass of the electron.

These relativistic effects are real. Need I go into other physics aspects of the material world which make no sense to our every experience of life? One such area is Quantum Physics, which states everything is quantized. For instance, energy only comes in discrete packets—one doesn't go from one energy level to another smoothly, continuously, but rather must somehow jump to the next level with no intermediate level. For instance, in everyday life, it would be as if one went in a car from fifty to sixty miles per hour in a single leap, and conversely in a single leap to slow down. That would seem to be absurd, yet in the atomic world, that is exactly what happens. Electrons surrounding an atom have only fixed levels of energy. If an electron moves to a higher energy level, it must leap to the next level. Sounds absurd, but don't just believe me, Dr. Richard Feynman, a Nobel Prize winning physicist, readily admitted, "Quantum Physics is absurd!" Yet it is true. Think about it. If an electron could just gradually lose energy, it would just spiral into the nucleus of an atom, but the fixed energy level of the electron's orbit keeps it from doing so. Quantum effects are so important to our material world, and thus to life. In fact, it is quantum effects on which our computers and solid-state electronics are based. How could this social network generation live without computers? From these examples, one must conclude, nothing, in its true essence, is completely what we in our everyday life experience think it to be.

Everything we believe about anything is but an approximation or estimate or glossed over picture of the true reality of what "something" really is. Science is not alone in telling us this. Religion tells us the same thing. Religion tells us that there is something transcendent to ourselves. There is a true reality which is beyond anything we can fully understand. Reality in its FULLNESS, it seems, can only be incompletely understood. The reality we understand can only be based upon the knowledge and perspective of the individual. No one truly understands the FULLNESS OF REALITY. On this, both Faith and Science agree. Maybe they are not so distant after all.

Let's take a simple example from everyday life which shows how one's perspective changes one's sense of reality. John and Mary were sitting at their maple, country-style kitchen table one night. Mary, a physicist, who was feeling philosophical that night, said to John. "Honey, what do you see when you look at this table?" John, an accountant, sensing that he was about to get into trouble, replied, "Dear, I know what you are about to say. I know I spilled some soup on the table, and I was just about to wipe it up." Mary began laughing and with a smirk responded, "John, I know you are a responsible man, and you certainly would clean up your mess before it would permanently stain my brand-new placemats! What I am asking is, what do you see when you look at this table?" Not feeling relieved, John looked intently at the table and said, "Dear, I see the large scratch I made when I put my tools on the table three months ago, and, yes, in addition, I know that the table has become wobbly and it's time to replace it." Frowning, Mary said, "No John, what do you see?" "Dear, all I really see is dollars, and you know we don't have the money right now." "John!" Mary said exasperated. "Well, Dear, what do you see?" John interjected fearing the worst.

"John, you look at that table and all you see is a table. How boring!" Mary, the physicist, replied. "When I look at that table, I see what it really is. It is almost nothing but a big void. The table looks solid to you, but, to me, the table is not solid, it mostly just empty. The atoms which comprise the table are, relative to their size, far apart from each other, much like the planets in the earth's solar system relative to their size, are distant to the sun. Similarly, the atoms in my finger, like those in the table, are separated by relatively large distances. My finger, like the table, is also mostly a void. One would think with such relatively large distances between atoms that if I pressed my finger onto the tabletop, my fingers would simply go through the tabletop. But our experiences tell us that this is not so. Instead, mysterious electric fields in the table arising from even tinier particles called electrons in the wood repel the electric fields arising from electrons in my finger. The electrons from my finger don't even touch the electrons in the wood. The fields keep them apart. The fields themselves are mysterious in the sense that the fields have no mass or anything, yet are able, when reacting to electric fields arising from other electrons, to create a force to repel the other electrons. While we know quite a bit about the characteristics of these fields—among these the strength, direction, and polarity of the field at any point in space and time—yet the details of why the fields create the force escape us. We can go even deeper into the physics and state the necessary conditions for subatomic particles which cause an electric force to arise, but, again, this does not answer the question why."

John looked at his dear wife Mary with his eyes glazed over, completely oblivious as to what had been said. Yet, he knew the reality of what had been said. So, he said, "Okay, Dear, get your coat on, and we'll go buy that new kitchen table you've been wanting." John then added, "And we'll buy some flowers for the new table."

Obviously, the reality of the kitchen table was very different for John and Mary. All John saw was regret and dollars. Mary saw the table in its true physicality, in a way not many would see, and, yet also saw a chance to get her new table she wanted. Both saw a true reality. Would others see things differently? What would a carpenter see, or an artist, or a child? What about Grandma, whose table that used to be, who sat her four children at that table with her now deceased husband? What would she see looking at that table? And, as for the flowers, how inadequate the interpretation would be if one simply looked at the narrow reality of flowers being but a collection of molecules.

Reality is not simply what our senses observe. There are non-physical messages and meanings attached to everything. The FULLNESS OF REALITY is so much more than a single view. One cannot explain the Universe, us, or anything else utilizing just a single view such as science, medicine, or even the Bible. The impression of these realities, as seen by the scientist/poet/artist/musician/child or even by animals, may be accessed using a physical medium, such as the brain, but are not generated by molecules. How could science alone accommodate all those realities? Without examining the total reality, one will never be able to determine the meaning of life. A strictly faith-based view of life has similar difficulties. It also is too narrow in its scope. In life, so many arguments arise between individuals because one is simply viewing the object or situation at hand from a different perspective. This is a major problem in reconciling science and faith. Without the inclusion of all aspects of reality, one will never approach the FULLNESS OF REALITY.

Mutual Willful Blindness

There are individuals and groups on both sides of the Science vs. Faith debacle that seem to be guilty of what C. S. Lewis calls, "willful blindness."[4,5] On the scientific atheism side, individuals, such as astrophysicist Stephen Hawking, evolutionary biologists Richard Dawkins and Jerry Coyne, etc., are all bright people who are, or have been, in the forefront at one time or other in the news because of the lack of belief in any spirituality. Because of such media prominence, many people believe that all scientists share their same views. That's just not true. There are many scientists who are also all bright people who do not believe in materialistic scientism—people such as Francis Collins, who headed the National Human Genome Research Institute, Nobel Prize winners Brian Kent Kobilka (Chemistry 2012), Peter Grunberg (Physics 2007), Charles Townes (Physics 1964), and Joseph Murray (Medicine 1990). For them, faith was/is an important part of their life. Like nearly all scientists prior to the rise of the Enlightenment in the late seventeenth and eighteenth centuries, science served as a glimpse into how God made things come about in such an amazing, complex, and beautiful way.

---

[4] C.S. Lewis, *Surprised by Joy* (San Francisco: Harper One, 2017), 256.

[5] Willful Blindness is also a recognized legal term.

On the Science Side

But what is true about the faith/science controversy is that the extremists seem to get the most press in America and elsewhere. Unquestionably, all these individuals are brilliant in their fields. However, on the scientific atheist side, many are unwilling to examine with an open mind evidence for the spiritual. For instance, Stephen Hawking, a world class astrophysicist, stated at the very beginning of his book entitled *The Grand Design*: "This book is rooted in scientific determinism, which implies that … there are no miracles, or exceptions to the laws of nature."[6] By starting off denying the possibility of a miracle to occur, Dr. Hawking was no longer an unbiased observer. In other words, Dr. Hawking simply blocked off consideration of data, not because the data was erroneous, but solely because the data did not fit his worldview. He preferentially chose one data set in his worldview. "Just close one's eyes and maybe the data will go away" must be the model.

Doesn't this remind you, in a way, what many medieval scientists did when they refused to believe theories that the Earth revolves around the Sun because it didn't fit their worldviews? Indeed, Dr. Hawking saw no need to evaluate other thoughts outside of science, including those of

---

[6] Stephen Hawking, and Leonard Mlodinow. *The Grand Design* (New York: Bantam Books, 2010), 34.

Philosophy, because for Dr. Hawking, "philosophy is dead."[7] For many atheistic scientists, scientific determinism is the predictor of everything. Some even claim that scientific determinism can predict everyday social outcomes.

So, what are some various forms of scientific atheism? Scientific atheism generally goes under the name of *Scientism.*[8] There are two major forms of *Scientism—Materialism* and *Reductionism.* In *Materialism,* one believes that only matter "matters." Only the things that can be detected by the senses or by some form of instrument constitute reality (of course, including those "things" which have no matter, such as light). In addition, one must include "things" which are not detected by instruments directly but are only detected because they affect something else in their environment. An example of this is the following: Jane, the physicist, heard a phone ring and answered it. She heard a sound coming from the phone from a man in another city. She began to cry. She really didn't hear him directly. Instead, she heard a sound which was detected by a microphone in the distant city. She heard the sound because the microphone in the distant phone converted sound to an electrical impulse which, through many steps, was transmitted to Jane's phone and re-

---

[7] Hawking and Mlodinow, 5.

[8] Baglow, Christopher *Faith Science and Reason—Theology on the Cutting Edge* (Woodridge IL: Midwest Theology Forum, 2012), is an excellent reference for those who want to go deeper.

converted back to sound through a speaker. You get the point. This is materialistic even though her senses didn't hear the actual sound that was transmitted through material objects. But the idea that only material activities are real is such a narrow view of reality. The real reality in Jane's phone call was that her relative was in a serious accident, not that sound waves hit a microphone. Reality extends beyond things.

For a true picture, all evidence should be used to evaluate the reality of every outcome. What if in the previous story about Jane's phone call, instead of a phone call, Jane had said she just had a feeling that something happened to her cousin Mike, then the phone rang? My wife sometimes gets these feelings. She can't explain it, but soon after such feelings, a phone call comes in, revealing this or that had taken place. Is my wife to deny that information, that data set, just because a material device such as a telephone wasn't used? Should I tell my wife that she is just acting in an unscientific manner?

The other branch of Scientism is Reductionism. In this thought, all of reality is reduced to its smallest physical part. For Carl Sagan, the noted American physicist, creator of the TV show *Cosmos,*[9] this meant that he, Carl Sagan, was nothing but a collection of atoms bearing the name "Carl

---

[9] Carl Sagan (November 9, 1934 – December 20, 1996), noted American Astrophysicist, co-writer of the award-winning 1980 television series entitled *Cosmos: A Personal Voyage*.

Sagan." One thing I am quite sure of, Carl Sagan's mother, who truly bore him—not just his name—would not have agreed that her son is just a collection of atoms. Her son was much more than that. She bore his atoms AND his personhood for nine months and raised him for many more. So, we see, one aspect of Carl Sagan is solely physical, and another is solely non-physical. The data set on reality must contain both physical and non-physical data. True reality goes beyond physicality.

So, where would we be if scientists had just ignored the data which led to the Theory of Relativity, Quantum Theory, String Theory because it did not fit the then-current concept of physics? Who would have thought in the 19th century that time, length, and mass can vary depending on how fast you are moving, or that only certain energy states are allowed in the electronic orbitals, or that movement is discrete rather than continuous? Who would have believed that there might be ten dimensions instead of the four commonly understood—that is, 3D plus time?

One must remain open but may be skeptical to the data at hand. Maybe the research into a reality which recognizes the physical and spiritual realities and merges them into a consistent picture of the fullness of reality will come to pass. I am sure there might be frustrations along the way, but that also happens when doing physical scientific research. Even dead ends are useful because now one knows one needs to develop a different approach. Willful blindness is always a mistake. Throughout this book, I will address some various examples of things in life which such scientists ignore.

On to the Religious Side

Enough about scientists and the like—at least for now. On the other side of reality, there are certain religious groups, such as Fundamentalists, some Evangelical groups, some non-denominational groups, etc., who believe in Scientific Creationism (the attempt to force Science to agree with the Genesis story of Creation) and fail through "willful blindness" to examine the evidence that science provides. They fail to see that Scripture itself points to science as a means to see the revelation of God. Paul in his letter to the Romans states, "*For what can be known about God is plain to them (the wicked and ungodly), because God has shown it to them. [20]Ever since the creation of the world his invisible nature, namely, his eternal power and deity, has been clearly perceived in the things that have been made. So they are without excuse*" (Romans 1:19-20[10]). Science studies the things that God has made. In studying science with an open mind, man can come to know about God. Science should lead one to God, not away from God.

---

[10] Unless otherwise stated, all Scripture quotations come from *The Holy Bible: Revised Standard Version* (New York: Collins, 1989).

Because of their literalist interpretation of scripture, many choose to discard the truth of Science. They do not understand that Truth reveals, not hides. Instead, many hold that the universe was created in six 24-hour days, that the Earth is just 6000 years old. While God certainly could have done all that in those time frames, there would be significant issues that would arise. God would have had to create the world to appear that it is much older than 6000 years. However, quite the contrary, God has revealed with ample evidence that He chose to allow the Universe to develop over 13.8 billion years, the Earth over 4.5 Billion years, Biological life over 3.9 Billion years, and *Homo sapiens* about 50,000 to 100,000 years.

It would seem that these religious organizations are afraid that their whole faith-based system would fall apart if the Bible were to be interpreted in other than a literalist way. Just because Creation did not occur in six days according to the scientific truths does not mean that the Bible is in error. The Catholic Church, from its earliest days, has pointed out that Genesis is not a science book and was never intended to be. For instance, St. Augustine wrote of the dangers of using Scripture to define science.[11]

---

[11]St. Augustine, *De acta cum Felice Manichaeo*, I, 10, (PL, 42.525).

"One does not read in the Gospel that Jesus said, 'I will send you the Paraclete who will teach you about the course of the sun and the moon.' For He willed to make them Christians, not mathematicians."

Furthermore, he believed what we now call Scientific Creationism would drive people away:[12]

It is a disgraceful and dangerous thing for unbelievers to hear a Christian, presumably giving the meaning of Scripture, talking nonsense [on scientific subjects]...If [unbelievers] find a Christian mistaken in a field which they know well and hear them maintaining his foolish opinions about our books, how are they going to believe those books in matters concerning the resurrection of the dead, the hope of eternal life, and the kingdom of heaven?"

In 2005, Cardinal Schönborn (Archbishop of Vienna) stated his concerns regarding scientific creationism:

The Catholic position is clear. St. Thomas (Aquinas) says that "one should not try to defend the Christian Faith with arguments that are so patently opposed to reason that the Faith is made to look ridiculous." It is simply nonsense to say that the world is only 6,000 years old. To prove this

[12]St. Augustine of Hippo, *De Genesis ad literam*, I,19.

> *scientifically is what St. Thomas calls provoking the irrisio infidelum—the scorn of unbelievers. It is not right to use such false arguments and to expose the Faith to the scorn of unbelievers."*[13]

The Catholic Church and others maintain that the Bible is to be interpreted in a literal way—as the inspired human author intended, not in a literalist way (i.e., the phrase "it's raining cats and dogs" means cats and dogs are falling from the sky). Instead of forcing science to match Scripture, or Scripture to match science, the true exegete of Scripture always tries to determine what it is that the author intended.[14] From there, one can seek additional spiritual meaning. Scripture is never intended to be a documentary. There is always a spiritual meaning. Scholars have shown that the Bible contains many genres of writing. A few examples are narration, poetry, prophecy, wisdom, biographies, and letters. In addition, there are many literary devices, such as similes, metaphors, irony, sarcasm, symbolism, typography, anthropomorphism, etc. If all the verses

---

[13] Christoph Cardinal Schönborn, "In the beginning God Created..." Available online at http://www.stephanscom.at/edw/kalechesen/articles/2005/12/02/a9719/

[14] For an excellent account of the proper way to read the Scripture, see the Second Vatican Council's Dogmatic Constitution on Divine Revelation (*Dei Verbum*).

written are to be interpreted in a literalist way, much of the meaning of Scripture would be missed.

In the literal view, Genesis 1 and 2, though speaking of true events, is written in a poetic language in order to explain Creation in terms of God's love and plan for mankind. If, indeed, Genesis had been written in a scientific factual way, then certainly no one would have understood it then, much less now. It makes sense that Genesis would be written in a way understandable to the people for whom it was written.

As I have previously stated, a FULLNESS OF REALITY can only be achieved by investigating all points of view, and a fullness of reality is necessary in order to obtain the FULLNESS OF TRUTH. After all, that is what both Science and Religions seek. As Truth cannot contradict Truth, we must realize that there may be different, incomplete views of Reality involved. With a greater understanding of these different views, we will find that both Science and Religion can be reconciled. Let's begin to start reconciling some of these differences between Faith and Science by looking at THE BEGINNING.

Things to Think About

1. Compare the literal versus the literalistic approaches to interpretation. What are a few of the literary genres found in the Scriptures?

2. If the first 11 chapters of Genesis are written in a mythopoetic fashion, how does that affect the meaning of those chapters? Does that mean the events described are not real?

3. Pick any object and construct at least four realities about it. Because there are physical descriptions of an object, does that mean the other realities are not real?

4. Can one determine "meaning" in anything if one restricts one's understanding to materialism?

Chapter 2

In the Beginning

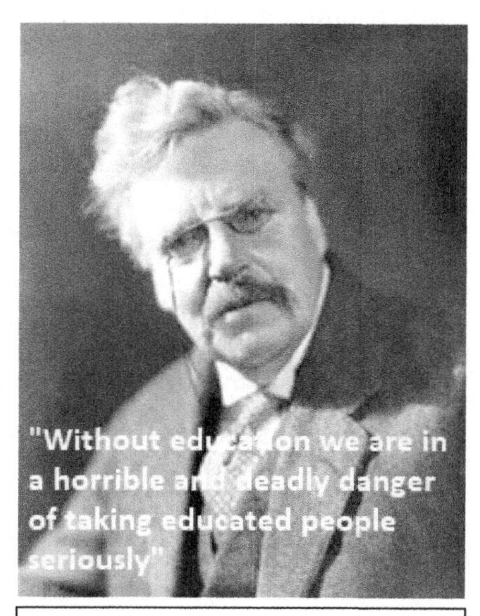

Figure 2.1 G.K. Chesterton, early 20th century English writer, philosopher, lay theologian, and literary and art critic

Besides the obvious difference as to whether God exists at all, there are two other major disagreements separating atheists and theists. The differences both surround the topic of origins. One disagreement involves the origin of the Universe. The other disagreement is the origin of life. In the religious point of view, the origin of matter and life are intertwined. One cannot speak of the origin of the Universe without there being somewhere in the back of one's mind the context of the origin of man. For the scientific atheist, the origin of the Universe does not, of course, require the origin of life as perhaps there could exist other universes without life, but, certainly, the origin of life does require the origin of the Universe. Obviously, there would be no arguing going on if intelligent life did not exist! But there is intelligent, rational life, so let's just continue on to the questions at hand. To discuss these origins, this chapter will cover the origin of the Universe with regards to the age of the Universe. Chapter 3 will deal the origin of the Universe as to the random nature of the Universe. Chapters 5 and 6 will deal with the origin of life.

The Religious Point of View

All cultures have religions and all religions, including the atheist religion (for atheists also have faith in science—just not in a supreme being), have addressed the issue of how the Universe began. All such origin stories speak of events which are pre-history. Those living in pre-history days told stories verbally attempting to try to explain their very existence (the who am I and why am I questions) while also relating to their progeny what had happened in their lives. Eventually, the stories were written down, which seems obvious but needs to be stated. Also, none of the pre-history origin stories are based upon science, which is not surprising since scientific studies and equipment necessary to support science simply did not exist. All of these stories, though, are consistent in the belief that someone greater than the storytellers created the world.

Many accounts involve "ugly" circumstances involving disagreements among the gods. In the Babylonian Creation myth, for example, a female god Omoroca is slain by the god Belus, who cut her in half, forming Heaven of one part and Earth of the other. The deity Belus later has another god cut off Belus' own head with the instruction to mix his blood with the Earth's soil in order to form man who is to become a servant to the gods.[15] Creation of man as servant is a common theme.

While most ancient stories of the creation of the Universe involve many gods, the story of the origin of the Universe found in both the Jewish Pentateuch (therefore in the Christian Bible) and the Muslim Quran depict a far different story. In the Jewish, Christian, and Muslim scriptures, the Universe was created by a God, not multiple gods. This is a major difference. In the monotheistic view, God (with a capital G) is all powerful, all knowing, and eternal. By definition, there cannot be more than one God, for if there were more than one god, then one of the gods would have to be less powerful/less knowing. Polytheistic gods on the other hand are just superior to humans, possessing more powers

---

[15] Cory, I.P., Richmond Hodges, E. (ed.), *Cory's Ancient Fragments of the Phoenician, Carthaginian, Babylonian, Egyptian and other authors* (London: Reeves, 1876).

than man, including typically, but not necessarily, immortality. To reflect this "all-everything" of Judaeo/Christian/Muslim thought, this god will now be spelled with a capital G. From here on, I will, for simplicity, discuss only the monotheistic God as polytheistic gods do not meet the criteria of god as all-everything.

According to Judaeo/Christian/Muslim thought, the Universe was created *ex nihilo*—out of nothing. It was a planned event and created in an orderly fashion. The Universe was created to be a home for man and necessarily would be anthropic, that is, capable of supporting human life. The evidence that the Universe was created to be anthropic (and thus was planned) will be presented in the next chapter. So, one can say that monotheistic religions agree that the Universe was created by God for man. In other words, the creation of the Universe was an intentional act of God, not a statistical event.

On the Age of the Universe

Let's now discuss the actual Creation event (whether by God or by accident). A major point of discussion is how old is the Universe? The importance of this is that the Judaeo-Christian scripture in the book of Genesis explicitly states a six-day time frame for the creation of the Universe to the creation of man while science describes an almost 14-billion-year timeframe. If this can't be resolved, either science or Genesis is wrong.

The media tends to present two views of this event—a fundamentalist Christian and an atheistic view—with seldom any attempt to reconcile the differences between the two sets of beliefs. In the fundamentalist view, God created the world out of nothing in six, presumably, twenty-four-hour days as stated in the Book of Genesis, and thus the universe is on the order of six to ten thousand years old. Meanwhile, in the atheistic materialistic view, the universe is just a random event arising out of a pre-existing, eternal medium which needs no creation as it has always existed. Some physicists believe that our Universe is the only universe; other physicists believe in the Multiverse view of our Universe creation. In the Multiverse theory, a parent universe caused our Universe to "pop" into existence, and we are but one universe out of almost an infinite number (more on this later). Neither physicist view can actually explain *why* this happened; each simply focuses on what happened after the "Creation event." Regardless of their approach to the creation event, nearly all physicists accept that our universe is about 13.8 billion years old (although the specific timeframe could change as more evidence is discovered).

The fundamentalist views and the physicist views are obviously completely opposite and, while we wish it weren't true, controversy sells newspapers, magazines, and advertising on the Internet. It is, thus, no wonder why the media projects such a clash. However, both of these views "need blinders" to be fully believed. Seldom do they mention Islam and 80% or so of Christian religions teach or at least allow that the creation of the Universe and man occurred in eons of unspecified length. It is then a misnomer for one to believe that all Christian religions believe in a six-day creation event. It is also a misnomer to believe all scientists believe that a truly random, unintentional explosion leading to the generation of our universe. Many physicists do accept the view that a supreme being is the cause of our existence and are happy to discover the physics of the event.

Fundamentalism

Let us examine the fundamentalist view, the problems the fundamentalist interpretation of scripture present, and how other monotheistic religions interpret the Genesis six-day account of Creation. Fundamentalists take the view that the Bible is to be read in a literalist fashion and therefore Creation occurred over a six-day, twenty-four-hour period. This creates problems and has resulted in Christianity being ridiculed as a religion by some scientists and those who trust in those scientists. Certainly, from one point of view, an almighty God could have created all the Universe, the Earth, animals, and man in a six-day period, but because He could do that does not mean the evidence shows He did it that way. So, let's look into some of the arguments for and against a six-day creation event, remembering from the introduction to this book, that Truth is Truth, and one truth cannot contradict another. There has to be a FULLNESS OF REALITY that accommodates both.

As a Christian, seeking Truth means one is seeking Jesus, for Jesus is "the way the Truth and the Life" (John 14:6). So, let's see what truth is out there. Up until the 18th century many, but certainly not all, believed in a young Earth age of around 6,000 to 10,000 years, but with the development of the scientific method that began to change. Using rock formation studies and eventually radiometric dating (evaluating the amount radioactive material in a known sample), the calculated age of the earth is now scientifically considered to be about 4.5 billion years old. Are those who believe otherwise, just well meaning, but uneducated? Perhaps some, but certainly not all. Many are highly educated, and some have a PhD in a scientific field. They are eager to point out the flaws in each of the date estimation methods. For instance, rock formation studies generally assume that the geologic layers were laid down slowly over long periods based on the rates at which we see them accumulating today. However, we now know that rock formation has often been associated with major catastrophic events which have resulted in huge upheavals of rock formations. This can invalidate the projected time calculations.

Likewise, Radiometric dating, while becoming more and more precise, can still contain large inaccuracies. The principle of Radiometry is that rocks and other materials will have isotopes. In theory, if one knows what the initial activity (concentration in non-scientific terms) of an isotope was at a particular time and then measures what the activity is now or conversely the amount of daughter atoms produced, one can calculate the amount of time that has elapsed.

Sounds good, but there are issues. First of all, one can't absolutely know what the initial activity was, and secondly, one doesn't know if there had been any radioactive contamination of the sample, which may have been added due to geologic events. An example of this issue was the Mount Saint Helen volcanic explosion in 1986. Radiometric dating in 1996 of samples of lava from the crater showed the age of the samples to be 350,000 years old instead of 10 years old.[16] Similarly, lava flows from Mount Ngauruhoe in New Zealand, which erupted in 1954, were determined to be 3.5 million years old rather than 50 years old.[17] One can see why the Creationists doubt the OLD age dating of the Earth!

---

[16] S. A. Austin, "Excess Argon within Mineral Concentrates from the New Dacite Lava Dome at Mount St. Helens Volcano," *Creation Ex Nihilo Technical Journal* 10.3 (1996): 335–343.

[17] A. A. Snelling, "The Cause of Anomalous Potassium-Argon 'Ages' for Recent Andesite Flows at Mt. Ngauruhoe, New Zealand, and the Implications for Potassium-Argon 'Dating,'" in *Proceedings of the Fourth International Conference on Creationism*, ed. R. E. Walsh (Pittsburgh: Creation Science Fellowship, 1998), 503–525.

In addition, other evidence suggests a much younger Earth. One example is that scientists have determined that the magnetic field about the Earth is decreasing at a known rate each year. Based upon the measured decrease in magnetic field strength, the magnetic field would have disappeared long ago if the Earth is truly 4.5 billion years old but would still be here if the Earth is just 10,000 years old. However, this concept has flaws as science has shown that the Earth's magnetic fields have had many changes over the years. At times, the Earth's magnetic fields have gone to zero strength, only to reappear later. The Earth's magnetic fields have even flipped, moving the North Magnetic Pole to the region of Antarctica, and, conversely, the South's Magnetic Pole, to the Arctic Circle. In 2009, it was noted that the position of the North Magnetic Pole is shifting over 34 miles per year toward Russia.[18] It is evident that changes in the magnetic field strength and positioning are not a good indicator of elapsed time and are more an indication of the change in the flow of molten iron and/or other paramagnetic materials in the Earth's core.

Another question mark is the observation that the moon is moving away from the Earth at a small but consistent rate of about 1.5 inches/yr or about 4 cm/yr. If the Earth is only 6,000 years old, then 6,000 years ago the Moon would have

---

[18] North Magnetic Pole, available online at en.wikipedia.org/wiki/North_Magnetic_Pole.

been only 800 ft (0.23 Kilometers) closer, a small change compared to the separation of the Earth from the Moon of about 240,000 miles (about 384,000 km). However, if the rates have remained the same, the Moon would have been touching the Earth only 1.4 billion years ago. As it goes, the Earth-Moon combination has been estimated to be 4 billion years old.

One last example (and there could be many more) is the existence of soft tissue found associated with dinosaur fossils. Young Earth Creationists state soft tissue could not have survived the 66 million or so years projected by scientists since the extinction of dinosaurs. Based on my life experiences, that certainly sounds reasonable, yet that hardly makes it true. In fact, scientists have discovered fossils of sponge embryos which date back over 600 million years.[19] For every tit there seems to be a tat.

It would appear that the Young Earth Creationists may have their reasons to believe in a young Earth and may not be WILLFULLY BLIND as many of the atheistic materialists think; however, alternatively, it may just be that there are flaws or inconsistencies in the systems for measuring the age of the Earth that have yet to be resolved. Perhaps bombard-

---

[19] Chen, Jun-Yuan et al. "Precambrian animal diversity: Putative phosphatized embryos from the Doushantuo Formation of China" *Proc Natl Acad Sci U S A.* 97(9) (2000) 4457–4462, at ncbi.nlm.nih.gov/pmc/articles/PMC18256/.

ments of the Earth by asteroids, orbital variations due to other planet motions, or whatever have caused the Earth to appear young.

So, is there other, much stronger evidence which contradicts a six-day Creation event and an age of the Universe—not just the Earth, which shows that the Universe is older than 10,000 years old, say 13.8 billion years old? That answer would have to be a strong YES. The expansion of the Universe, as predicted by Fr. Georges Lemaître and others at the turn of the 20th century and measured by Edwin Hubble in 1929, definitively shows that the Universe began as point and is growing larger every day. As far as we know, the Universe began "ex nihilo," that is, out of nothing. There is no evidence that anything, including time, ever existed prior to the moment of what originally was derisively called the "Big Bang" by physicist Fred Hoyle in 1949. More recently, some physicists have theorized that our Universe was generated by a mother universe (or whatever) and that there are multiple other universes all in their own "space-time," none of which can be detected because we are constrained to this universe in our measurements. Some theories maintain that our Universe arose as a blip within a much larger universe—that is, a sudden expansion of a particular region. Other theories hold that a "universe-generator" creates a new universe. These theories are referred to as the Multiverse.

Whether true or not is simply conjecture at this point and belongs to the world of theory and to mathematics. Why conjecture a Multiverse? Well, why not? That is part of science to postulate solutions to evidence which does not follow our existing data. But you might say, "You've already stated that we are constrained to measurements from within our universe, so how could data from outside our universe solve problems found within our universe?" Well, the biggest problem is that nearly all physicists believe that our universe is so finely tuned that everything has to be exact, or we wouldn't even have a universe, much less an anthropic one which allows for human life.

How exact? Well, as we will see in the next chapter at least 1 in 10 to the 10 to the 123 power. 1 in 10 to the 10 to the 2nd power is the same as 1 in 10 to the 100 power, or in more common terms 1 in a 10,000 trillion trillion trillion trillion trillion trillion trillion trillion trillion trillion trillion. Did I say common terms? 1 in 10 to the 10 to the 123 power can only be described non-sacrilegiously, as I heard from a pastor one day, as 1 in a godzillion!!!! In other words, the likelihood that the Universe was created as it was, is for all intents and purposes, totally impossible—yet it was created. Hmmm! So, in order to make it likely that our universe could exist, it has to be suggested that there may be a godzillion universes out there. Furthermore, if there is a mother universe, perhaps we don't need a God to begin it all.

Alas, however, some physicist mathematicians have shown that the parent universe and all other universes would also have to have order and a beginning.

Getting back to the main point as to which theory—Old versus Young Creationism has more data behind it. There are thousands of measurements and models all based on a Big Bang, which correctly predict the measurable aftermath of such an event. The data show that we are in a relatively young universe at about 13.8 billion years. The Universe will keep expanding, and, in fact, its expansion is increasing. Eventually, after more than 100 billion years as predicted by many physicists, the universe will expand so much that stars will be so far away that the night sky will be completely absent of light. Eventually, whatever stars remain will be burnt out. The Universe will "die out."

To believe in a Young Earth Creation, at this point, with so much consistent measurements matching theoretical predictions all based on an expanding universe in the aftermath of a Big Bang, is to deny Truth of science. Furthermore, it is equivalent to say all the scientists who have evaluated such measurements have made errors in their analyses or that they are deceitful liars. One could reconcile this by saying that for God's own reasons, God created the Universe to appear as if it were billions of years old. But as we know, "God is not man, that he should lie." (Numbers 26:19) and Jesus is "the way, the TRUTH, and the life." (John 14:6)

Yes, the Young Earth Creationists have pointed out that it appears there are some issues in measuring the Earth's age, but those, I believe, will be resolved in time and are much less significant than the expansion of the Universe. The question is not "What do you trust more—the measurements of thousands of highly educated scientists or the Scriptures?" but "Can the FULLNESS OF REALITY reconcile these differences?"

To begin reconciling these differences, recognize that it is the interpretation of the words of Scripture as intended by the human authors under the inspiration of the Holy Spirit that is inerrant. For this, we must look to see how exegetical methods determine the meaning of Scripture. To start out, one should easily see that the Scriptures are not science books. Therefore, we are not to read Scripture in a literalist fashion when speaking of matters of science. Scripture, after all, is intended to reveal truths about the WHO and the WHY of Creation, but not the specifics of the WHAT and the HOW. Let's examine the account of the creation of an anthropic Universe and Earth as found in Genesis 1.

Day 1 (Genesis 1:1-5)

[1]In the beginning when God created the heavens and the earth, [2]the earth was a formless void and darkness covered the face of the deep, while a wind from God swept over the face of the waters. [3]Then God said, "Let there be light"; and there was

light. [4]And God saw that the light was good; and God separated the light from the darkness. [5]God called the light Day, and the darkness he called Night. And there was evening and there was morning, the first day.

From the first verses, it is evident that the description of Creation is a poetic one. The writer states that darkness covered the face of the deep and a wind covered the face of the waters. "Face of the deep" and "face of the waters" can only be poetic expressions. God then separates "light" from the "darkness." We know that darkness is nothing but the absence of light. Thus, we have another poetic expression, yet we clearly understand that God is creating the universe and is acting with intent. However, one could argue that, as we will see in the next chapter, initially after creation, there was a period in which there was no light at all. Light appeared in about 100,000 years. God defines a period as a day, even though there is no day as we understand day. The human author is clearly indicating that creation was a process, not performed in an instant. There is no real scientific construct here, other than there was a beginning (Big Bang). Genesis doesn't state how the Universe physically started, only that it did. Physics, likewise, doesn't state how it started, but can explain only what happened after it started.

Day 2 (Genesis 1:6-8):

[6]And God said, "Let there be a dome in the midst of the waters, and let it separate the waters from the waters." [7]So God made the dome and separated the waters that were under the dome from the waters that were above the dome. And it was so [8]God called the dome Sky. And there was evening and there was morning, the second day.

On day two, the sky is created, and waters are present. The human author is simply indicating that the God is proceeding as intended, in an orderly process. Science acknowledges order in Creation.

Day 3 (Genesis 1:9-13)

[9]And God said, "Let the waters under the sky be gathered together into one place, and let the dry land appear." And it was so. [10]God called the dry land Earth, and the waters that were gathered together he called Seas. And God saw that it was good. [11]Then God said, "Let the earth put forth vegetation: plants yielding seed, and fruit trees of every kind on earth that bear fruit with the seed in it." And it was so. [12]The earth brought forth vegetation: plants yielding seed of every kind, and trees of every kind bearing fruit with the seed in it. And God saw that it was good. [13]And there was evening and there was morning, the third day.

In the poetic expression, on day three, dry land and the seas are created. Plants proliferate on the ground. This is all without a sun. All is intentional and orderly.

Days 4 to 6 (Genesis1:14-15, 19-20, 23-26, 31)

[14]And God said, "Let there be lights in the dome of the sky to separate the day from the night; and let them be for signs and for seasons and for days and years, [15]and let them be lights in the dome of the sky to give light upon the earth." And it was so.... [19]And there was evening and there was morning, the fourth day.

[20]And God said, "Let the waters bring forth swarms of living creatures, and let birds fly above the earth across the dome of the sky... [23]And there was evening and there was morning, the fifth day.

[24]And God said, "Let the earth bring forth living creatures of every kind: cattle and creeping things and wild animals of the earth of every kind." And it was so. [25]God made the wild animals of the earth of every kind, and the cattle of every kind, and everything that creeps upon the ground of every kind. And God saw that it was good.

[26]Then God said, "Let us make humankind in our image, according to our likeness; and let them have dominion over the fish of the sea, and over the birds of the air, and over the cattle, and over all the wild animals of the earth, and over every creeping thing that creeps upon the earth."

[31]God saw everything that he had made, and indeed, it was very good. And there was evening and there was morning, the sixth day.

Day 4 completes day 1, as the stars, sun, and moon fill the heavens. Day 5 completes day 2, as the birds fill the sky and the fishes fill the water. Day 6 completes day 3, as the dry land is filled with all kinds of living creatures, including man. Days 4 to 6 completing days 1 to 3 respectively is a poetic way of expressing that God created everything and gave them their rightful place. Again, everything is intentional and orderly.

However, here we see that the description of the creation of man is different from that of the other animals. Man is to be intentionally different from the animals. We are to be created in God's image, meaning we will have some of God's attributes that the animals will not have. We will have free will, which is necessary so that we can love, so we can choose to do right or wrong. Mankind will be responsible for and have dominion over the plants and all living creatures.

Stephen Hawking in his earlier days expressed the conundrum that atheistic materialists find themselves in:

> "*Physics is just a set of rules and equations. What is it that breathes fire into the equations and makes a universe for them to describe? The usual approach of science of constructing a mathematical model cannot answer the*

question of the question of WHY there should be a universe for the model to describe."[20]

Let me compare the writings of the earthly writers in Genesis to the writings of scientists concerning the substance of the Universe. Genesis simply uses the broad terms heavens and earth while science uses terms such as protons, neutrons, electrons, nuclei, plus even harder to understand terms such as quarks, bosons, pi mesons, etc., which few people, of course, have the foggiest notion about. Articles about Science which discuss the atom often contain the picture of the nucleus with protons, neutrons, and electrons similar to that seen in Figure 2.2.

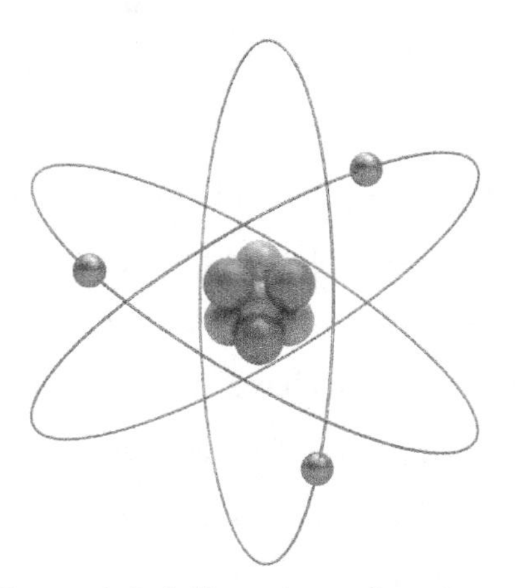

Figure 2.2 A Drawing of an Atom

---

[20] Stephen Hawking, *A Brief History of Time* (New York: Bantam Books, 1988).

But, do you really think the protons are one color, neutrons are a different color, and electrons are a third color? Of course not! Are each of these different particles truly spheres? Of course not! Do electrons travel in such well-defined orbits. Of course not! Does that make the illustration wrong? No! Scientists and educators have used models and pictures that really aren't true in order to depict their version of reality. The full reality is so much more complex, and no one can describe in drawings and words absolutely what is going on. And, of course, only a relatively small handful of people even begin to understand it. But the scientists and educators are not lying to us. They are simply trying to express truths concerning the WHAT and the HOW in terms we might understand. Similarly, the Scriptures, without specifying the science involved (which no one would have understood anyway, do you?), have managed to tell us the WHO and the WHY and a very unspecific HOW—that is, for example, God intentionally created from nothing a universe in an orderly way for us. As a result, the human authors of Genesis have revealed TRUTH about real events that did happen, and that was a very good thing to do. So, HOW did God create? That is not even the question. WHO created and WHY? Now those are questions theologians want to answer.

With the understanding that truth should be expressed in a manner that best answers the question actually asked, our Faith versus Science conflict can be resolved. Science simply tries to tell us WHAT the Universe is and HOW it was formed. Science cannot answer the WHO and the WHY. Similarly, as previously stated, religion answers the WHO and the WHY, but since the WHAT and the HOW are not their intended goal, answers in that arena are not specific in a scientific sense. So, it is not hard to see that neither the Scientific nor the Theological approach gives a FULL picture of Reality. Each Truth whether that be Science or Faith is but a PARTIAL Truth leading to a false sense of the TOTAL Reality. The actual union of the truths of Faith and Science becomes obscured.

However, Religion, one must admit, does offer a very undetailed response to the HOW of creation. Religion states that the Universe was intentionally created by an intelligent being out of nothing. Creation was caused and was not random. What does Science say about that? On to the next chapter where we will look at some scientific evidence concerning the origin of the Universe and see if reality supports the God version or the random version, or somehow, maybe both! For this, we shall look at the question of intentionality in the creation of our Universe.

Things to Think About

1. How would you explain the origin of the Universe, the Earth, and the Earth's plants and animals to an audience who knows nothing of science? Your ex-

planation must explain not only the How, but the Why.

2. What would be the purpose of describing science to such an audience? What is the real purpose for which the Scriptures were written anyway?

3. Many evangelicals/fundamentalists get around the six 24-hour day quandary by acknowledging that the word for day in the Hebrew scriptures “yôm” can also be translated as an epoch or an eon. Thus, they can maintain a literal inerrancy of the Bible by saying, “Well, it could have been a long time.” How does this approach still misunderstand the Inerrancy of Scripture? How does it miss the point of Scripture?

Chapter 3
What a Start!

In the last chapter, I indicated that the Book of Genesis properly shows that our Universe was created with Intentionality and Order. Scripture shows that the Intentionality and Order came from God. Physicists may not like the word Intentionality, but they don't disagree with it. They simply choose a different name. They call it Fine-tuning. Physics has no problem with the Order part. As I will discuss, the Universe has great Order to it. They call that Low Entropy—a term to be explained later. The dance that atheistic physicists have to make is to allow for both Fine-tuning and Low Entropy to occur without something causing it. Atheistic physicists and, therefore, atheistic scientists and other atheists who jump on board because they trust the work of the physicists (some of whom believe in God) might name this cause Probability while theistic scientists and similarly, of course, others who believe in the Transcendent, might name this cause God. It does depend on where your starting point is in your beliefs.

So, you may ask, what is Fine-tuning? Now, it is reasonable to assume that a wide range of people are reading this material with different science backgrounds. Bear with me, both the technically-minded and those with a less scientific background. I will try and make this as short and painless as possible for both groups. To answer, I must first explain the creation of time. Time is a measurement of change. Before the Big Bang, there was nothing. If nothing existed, there wasn't anything to change, so time didn't exist. With the Big Bang, energy was dispensed, and matter was or soon would be created. Now changes could be measured, and time began.

As the Universe began to develop, there were problems to overcome if the Universe were to become a place suitable for man. At 10^{-35} seconds (far shorter than a millisecond, a microsecond, a nanosecond, a picosecond or anything else you have ever heard of), a correction was made to the initial expansion of the Universe. The Universe suddenly inflated for the period between 10^{-35} seconds to 10^{-34} seconds. During that period, the Universe expanded a trillion, trillion times. Lewis and Barnes in their book *A Fortunate Universe* compare this expansion to a grain of sand expanding to the size of our galaxy—more than 100,000 light years across.[21] Wow! These numbers are just incredible, aren't they? The benefit

---

[21] Geraint F. Lewis, and Luke A. Barnes. *A Fortunate Universe* (Cambridge University Press, 2016), 170-171.

of this is that any "lumps and bumps" in the initial energy distribution of the Big Bang would be smoothed out, just as blowing on a balloon quickly smooths out the wrinkles of an un-blown balloon. Without this sudden inflation and smoothing, the Universe would have collapsed on itself or have expanded too quickly. How do physicists know this for sure? They don't! They simply know that it must have occurred if we were going to have the finely tuned Universe in which we find ourselves. Therefore, they have theorized it is an effect, not a cause.[22] Maybe they will find a reason why it occurred someday. Maybe not.

What happens next? Shortly after $T = 10^{-33}$ seconds, various subatomic particles with their specific attributes were created, which when combined in a certain manner create force fields. Typically, we name four types of forces, but that doesn't mean there couldn't have been more. One works with what one can know. The most common forces we are aware of are Gravity and Electromagnetism. Gravity is the force of attraction between masses, and Electro-magnetism is the force between charges or the result of moving charges. Thus, there is a gravitational force of attraction between the Earth and the Moon, the Earth and the Sun, and the Earth and your body which you note every day, especially when you step on the bathroom scale!

---

[22] Lewis and Barnes, *A Fortunate Universe*, 170-171.

You probably feel pretty comfortable thinking about the concept of matter as you look at all the objects around you, but you have been fooled. There are other forms of matter interacting with gravitational forces in the physics world, such as Dark Matter, which you can't see directly but which comprises 27% of the matter in the Universe, and Dark Energy, which comprises 68% of the universe.[23] (The percentages will depend on which physicist you talk to.) Surprise, surprise! The everyday matter we know and love (until you step on that bathroom scale) comprises only 5% of the universe's matter. So, what is Dark Matter and Dark Energy? I'd say, "Don't ask" because physicists are not exactly sure. Besides my head is spinning, and I'm getting a headache at this point. However, while physicists cannot directly detect Dark Matter and Dark Energy, they see unexpected gravitational effects. As a result, they proposed that there must be Dark Matter causing the gravitational effects. Dark Matter, we are told, is actually everywhere, including in the room where you are reading this book! Dark Energy relates to an energy which is causing the Universe to expand faster than it's supposed to and will cause the Universe to never return to a single point. The source of Dark Energy and what it exactly is, is unknown. Since good ol'

---

[23]Wikipedia, "Dark Energy," available at en.wikipedia.org/wiki/Dark_energy.

Albert told us $E=MC^2$, we can relate that energy to an equivalent mass.

Now for Electromagnetism. Subatomic particles can combine to make different kinds of particles. The most common particles that one is typically aware of are protons, electrons, and neutrons. Protons have a positive charge (whatever that is) and electrons have an equal and opposite negative charge. Like charges repel each other; opposite charges attract each other. Of interest, this has nothing to do with the mass of the particle. A proton with a charge of positive 1 has a mass about 1800 times greater than that of an electron, which has a charge of negative 1. A neutron, on the other hand, has zero charge and weighs just slightly more than a proton. And yes, the mass of a proton plus an electron is about the same as the mass of a neutron. Anyway, we have all felt the effect of the movement of charges. One common example, which we all learned as kids, was to walk shoeless on a carpet, then come up from behind and touch the earlobe of a friend or, better yet, a sibling. Zap! A spark went from your finger to your friend's earlobe. Such a cruel way to teach Science.

The other two forces are the Strong and the Weak nuclear forces. In an atom, as you already know, the protons and neutrons reside in a nucleus. Since the protons are so close together in the nucleus, there is a strong repulsive force that attempts to push them apart. Therefore, there needs to be a strong force to overcome the repulsion of the charged protons. Cleverly, this was named the Strong Force. In addition, there is another much weaker force which tweaks the forces in the nucleus and is associated with such events as the radioactive release of the nucleus particles. Again, cleverly, this has been named the Weak Force.

Enough of the description about the different forces. Here is what is important. Each force has a constant associated with it. The constant was defined at Creation and as far as anyone is able to tell is quite arbitrary. Let's say, just picking a number, the value of a constant for the force of Gravity is the number 1.10. If the value instead was 2.20, then everyone would weigh twice as much. If another universe exists, the value for the Gravitational constant might be 9.83—who knows, as there has not been established any law which would require a universe to have a certain number for a constant. Once established, physicists believe the constant is fixed as long as the universe exists.

So, why is this called Fine-tuning? In the old style of radios, you had to rotate the frequency dial so as to find your favorite radio station. If you didn't get it exactly, all you would hear would be electrical noise. Same is true of the Origin of the Universe event. If everything isn't exact, one doesn't get an anthropic universe with stars, galaxies, etc., such as we have. Here are some examples. It's Fine-tuning because if you change the value of the Gravitational constant by only one in a 100 billion billion billion billion billion

billion billion billion (or 1 in 10^{50})[24], then there would be no universe! This is an incredibly small number! If the change makes the force weaker, then the universe would expand too quickly, and the galaxies would not form. If stronger, then all the matter would condense back to a super black hole and again, no galaxies would form.

But Gravity is not the only force that is Fine-tuned. If the Strong force were 2% stronger, no Hydrogen would exist; if 2% weaker, only Hydrogen and Helium would exist. Either way, one can't have water, which is two hydrogen atoms and one Oxygen atom. Changes in the Weak force of only 1 part in 10^{50} (same as probability as a change in the gravitational constant) would have an effect on the production of heavier elements and the rate at which the stars burn. Similarly, if the Electromagnetic constants were stronger or weaker by 1 part in 10^{40} relative to the Gravitational force, the full range of star sizes and types necessary to make life possible would not form.[25]

Forces are not the only things that must be Fine-tuned. There are many other constants which must be established just right at the beginning of time, that is at Time= 0.

A few examples related to Space and Time are:

---

[24] Paul Davies, *The Accidental Universe* (New York: Cambridge University Press, 1982), 107.

[25] Hugh Ross, *The Creator and the Cosmos* (Covina, CA: 4th edition, RTB Press, 2018), 175.

Smallest distance interval possible 1.62 x 10^{-33} cm

Smallest time interval possible 5.39 x 10^{-44} sec

Speed of light = 3.00 x 10^{5} km/sec (186000 mi/sec)

Initial Entropy (to be discussed later)

Cosmological constant (rate of expansion of the Universe) < 10^{-53} in SI units

A few examples relating to the properties of matter are:

Rest mass of an electron (the mass when the electron is not moving) = 9.11 x 10^{-31} kg

Rest mass of a proton = 1.67 x 10^{-27} kg

The unit of proton and electron charge = 1.6 x 10^{-19} coulombs

Most people have never thought about how all these things must fit exactly right. If you change any of these, the Universe—that is, an anthropic universe—cannot exist! You might ask yourself, "Is it reasonable to believe that all these constants and values just happened to fit perfectly together when the tolerance is so small?" How could that be?

To top it off, let me give you one more example of an initial condition which is incredibly, incredibly unlikely and which was discovered by atheist physicists! This is the concept of the initial Entropy of the Universe.

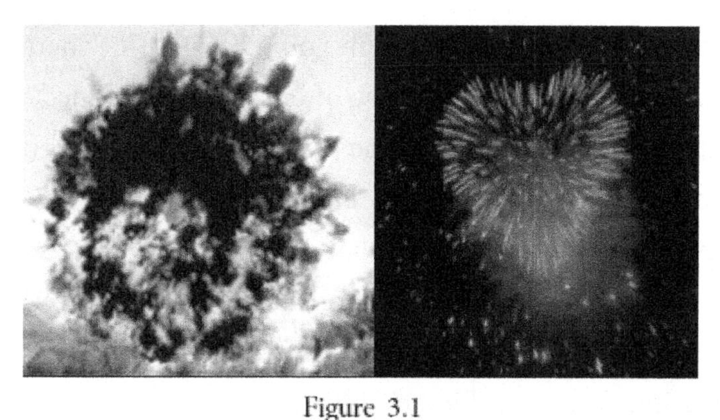

Figure 3.1
(a) A Bomb Explosion (b) A Fireworks Explosion

A Necessary Question - Who made these laws anyway or do they just happen?

Who enforces these laws, or are they self-enforcing? No one knows. Science just describes what appears to be physical laws. Need the Laws be the same in all universes, if indeed there is more than one universe?

Most physicists say NO.

So, bear with me just a little bit longer. Let's discuss Entropy. The Second Law of Thermodynamics states that all interactions result in the loss of the ability to do work and that order goes to disorder. This characterization of the loss of ability to do work is termed Entropy. Entropy measures the amount of decay or disorganization in a system as the system moves continually from order to chaos (sort of like my desk!). The LESS work one can do, the HIGHER the Entropy. I know, this sounds confusing, so let me show you an example. Figure 3-1 shows two explosions. Figure 3-1(a) is an explosion due to a bomb going off. Lots of energy is released. However, the release of energy is unplanned, disorderly and hardly useful. The explosion in Figure 3-1(a) is said to have HIGH Entropy. Figure 3-1(b) depicts another explosion. This explosion is that of a fireworks display. Like the first explosion, a lot of energy is released. However, the release of energy is obviously intended and orderly. A beautiful work is achieved. This explosion is said to have LOW Entropy. The first explosion, I would call a big bang (but not the Big Bang). The second, I think, is more appropriately called a Grand Launch. Is the second explosion fine-tuned? Yes, but not nearly fine-tuned to the same degree as the Origin of the Universe (the Big Bang).

Just how finetuned was the Big Bang? In 1989, Roger Penrose, an atheist physicist and associate of the well-known, end-of-life atheist physicist Stephen Hawking, calculated the Entropy of the Universe (an underestimate) and determined that the likelihood of our Universe having an such a low initial Entropy at Time = 0 is 1 in $10^{1000000000000000000000000000000000000000000}$ . . . where the zeros continue until you have 123 of them.[26] The number of years of our Universe's existence is about 13.8 x 10^9. It only has 9 zeros! One trillion is 10^{12} with twelve zeros. But let's not worry

---

[26] Roger Penrose, *The Emperor's New Mind: Concerning Computers, Minds, and the Laws of Physics* (Oxford: Oxford University Press, 2002), 339-345.

about numbers with such big powers of 10 because these numbers are so big that no one can really understand just how large these numbers are. As in chapter 2, let's just call the likelihood of the Universe having such a low Entropy is, in an appropriately pious way, 1 in a godzillion! In other words, it would appear to be completely unlikely.

Of course, the story of the creation of the Universe so that it is suitable for life, particularly for the life of mankind, hardly stops with fine-tuning of the Universe at T=0. Many unlikely things had to occur until the Universe, particularly the Earth, was made suitable for the establishment of life (Cosmic Evolution), to be followed by the initiation of life from inorganic materials (Chemical Evolution), and later, the establishment of our ecosystem and our particular human life form (Biological Evolution). Let's now look at the development of the Cosmos so that an anthropic Earth could be established. Like the Fine-tuning of the Universe at T=0, Cosmic Evolution is an extremely unlikely event if viewed as a completely random series of events.

Things to Think About

1. Can the development of the Universe be both Random and Orderly?

2. In Ancient Greece, Aristotle and a philosopher named Zeno had a disagreement about whether

motion was continuous or discrete (that is in steps). Zeno said discrete, Aristotle said continuous. Basically, in my version of the argument, Zeno would have said that if a runner moved 4 feet in one step, then half that distance (2') in the next, then half that distance (1') in the next, etc., he would never cross the finish line. Aristotle would have said motion must be continuous as the runner obviously does cross the finish line, so Zeno you are one crazy philosopher! Aristotle won the minds at that time, but he was wrong. How does a minimum distance interval show that Zeno was right?

3. There are so many requirements that force strengths, charge ratios, mass size, etc., must be exactly as they are if man is to exist. Some say that we were just lucky; others say we weren't just lucky, but that this happened, and, therefore, it must be achievable. So why worry about it? Why doesn't this answer the real question?

Chapter 4
Following Through!

In the last chapter, I introduced the concept of a godzillion—a number inconceivably so large as to practically be infinite. Fortunately, in this chapter the numbers and corresponding odds are a little bit better. When I was young, any large number was a gazillion. A gazillion was way, way past a trillion or even a zillion (I am not making this up, you can even find this in Wikipedia!), so the odds in this chapter will end up being smaller than a godzillion—only 1 in a gazillion. Still the likelihood is exceedingly small, and it is not reasonable to think they could actually happen on time, in the right amount, in the right order.

The Earliest Eons

So, what are some of the necessary steps that necessarily had to happen in a "Goldilocks" fashion—not too much, not too little, at the right time, at the right sequence, etc. Certainly, I will not pretend to cover all things, as exciting as it may be to me, because it might eventually put you to sleep! We left off in chapter 3 with all the laws and constants, energy, field strengths, etc., being initiated exactly right. But what happens next? Does one see a great light and hear a loud bang? Hearing requires air, and there is no air. Atoms must first form. After the Big Bang event, there were elementary particles, called Quarks, which must be fused together to form protons and neutrons. As the temperature cooled a tiny bit (however, it is still really, really hot!), negatively charged electrons were able to be "captured in orbit" by protons to form Hydrogen atoms (1 proton and one electron in "orbit"). Eventually, the temperature further cooled off and then some Hydrogen fused together (as in a Hydrogen bomb) to make helium (2 protons, 2 neutrons in the nucleus and 2 electrons in "orbit"). Later, some Lithium was added (3 protons, 3 neutrons and 3 electrons in "orbit"). If there had been too little energy or too much energy in the creation, then creation would not have happened, at least not in the perfect quantities which were needed to form the Universe that we know and love.

As each element gets created, it gets harder to create the next. Eventually, the created particles formed into gas clusters, and then into stars. There they would have had the necessary energy to create elements with greater and greater numbers of protons and neutrons. But, as stated, not all atoms are created with the same ease. We know that in the final product in an anthropic universe, we must preferentially end up with lots of Carbon, as all of Earth's biological life is a Carbon-based system. Why should Carbon have been preferentially created?

This question and its answer led to the conversion from atheism of the renowned atheistic physicist Dr. Fred Hoyle, who acknowledged that the Universe was designed. In his research, he found that, just like a suspension bridge has certain preferential frequencies (that is, stepping on a bridge rhythmically causes it to easily bounce up and down wildly, which is called resonance, while not doing so rhythmically results in the bridge not doing much of anything), carbon has the perfect properties with an exact resonance point so that Carbon is preferentially created over other elements enabling the mass production needed for a Carbon-based biology for Earth! But Dr. Hoyle didn't see it as a coincidence, and, in awe, stated,

> *"Would you not say to yourself, 'Some super-calculating intellect' must have designed the properties of the carbon atom, otherwise the chance of my finding such an atom through the blind forces of nature would be utterly minuscule. A commonsense interpretation of the facts suggests that a super intellect has monkeyed with physics, as well as with chemistry and biology, and that there are no blind forces worth speaking about in nature. The numbers*

one calculates from the facts seem to me so overwhelming as to put this conclusion almost beyond question."[27]

This was a statement coming from a man who worked with the vastness of the Universe most of his professional life. One can see how Science can lead one to faith; however, in this case, Hoyle still refused to call the "super-calculating intellect" God!

Days are Created

Elements eventually gathered into gases, gases into stars, stars into galaxies, and clusters of stars into Local Groups. Gradually gases and cosmic debris clumped together into planets—some solid, some gaseous—and began orbiting their nearby star. Voilà! Our solar system was formed with its central sun surrounded by planets. Furthermore, as our earth is rotating about its axis, we now can talk about as a "day," and the time it takes to orbit the sun, as a "year." Sounds simple doesn't it? Believe me, it wasn't as simple as that. It is simple only if one doesn't examine what has to happen for all this to occur. And, of course, no one person can examine all that has to happen. We all know that the Earth revolves around the sun, for instance, not because we

---

[4-27] Fred Hoyle, "The Universe: Past and Present Reflections." *Engineering and Science*, (1981), 8–12.

really know such things, but because a relatively small number of people have been in the position to measure such events. Was it really unreasonable, when one is standing still on the Earth, gazing into the night sky, to think that the heavens are moving and we are not? The rest of us know this now only because we trust others.

So, what had to happen for the Earth to be as it is today? Did the events I will soon describe happen because they were forced to by the forces and associated force strengths created at the dawn of the Universe, or did they just happen as a series of random events, and we got incredibly lucky? Or could it be possible that a "*super-calculating intellect*" didn't just create the universe then go out the back door, but rather has had a guiding hand in the Universe at all times, even today? Let me discuss three possibilities.

The first possibility to consider is that, following Hoyle's lead, the Universe was designed by God or some Super Intellect. Once designed, the designer, just stepped aside and let things happen. I am not necessarily even claiming a divine design, but just the considering the Universe with all its initial constraints—energy employed, forces generated, constants employed, etc., as a design set. Could an initial design account for the series of events that happened, resulting in our anthropic Universe? No, that could not be the case. This can easily be established. "What?" You might ask. "Even if God, or Hoyle's Super Intellect, set the perfect design constraints at T=0, one couldn't say that our Universe and our

Earth, would-end up being the wonderful creation we have today?" Yes, that is exactly what I am saying. Even a perfect design would not necessarily result in a final perfect universe. The reason is that physics has discovered that all interactions are determined by probability. This is reflected in such concepts as the Heisenberg uncertainty principle which states, for instance, that we can't know at any one time both the position and momentum (direction and kinetic energy) of a particle even if the initial conditions of a particle could be known. In other words, what happens, happens within certain limits. Statistically, some things are more likely to happen, but that doesn't mean they will happen. Thus, even if another universe existed with the same forces, constants, etc., the likelihood of another universe being exactly the same as ours is impossible, unless of course, one wears the rosiest of rose-colored glasses. Well, maybe, it doesn't have to be the same, just close. I will go along with that; however, as shown in chapter 3, the margin of error is so small that one might as well say it has to be exact.

As far as a second option, let's just say for now that everything happened by random probabilistic events, and we got what we got. We are just so lucky. Atheistic materialists believe that is what happened. And what about the third option? That is called religion. The monotheistic religions do not believe that God created everything and then left us to fend for ourselves. Instead, they believe that if God were to remove Himself, then all of the Universe would dissipate. In some ways, religion is not much different from science. Quantum Physics theorists also maintain a similar concept. They theorize that a particle is not a particle unless it is observed. For religions, that observer is God.

A List of Events

Let's return to option 2—the totally random possibility. In this situation, things changed somewhat randomly over a given period, which makes it an evolution. In this case, I am speaking of a Cosmic Evolution, not to be confused with the two other great evolutions: the Chemical Evolution, that is the development of molecules which could have led to life, or the Biological Evolution, which proceeds from the first cells to the current stage of living things. In constructing a list of some, but not all, events of cosmic evolution, one has to first understand that much of this is scientific conjecture. Much of the conjecture occurs by looking at present conditions, seeing what we have now, then proposing models that will result in a planet's being like what we have. Not everyone will agree with what I have compiled, and, furthermore, what is stated will most likely be changed, or at least nuanced, in the upcoming years as more knowledge is garnered. Nonetheless, our luckiness in having the Earth turn out the way it has should certainly be appreciated.

First of all, we are lucky enough to be in a galaxy such as the Milky Way. The Milky Way Galaxy (MWG) is a spiral galaxy and is very stable over its long time period. Stability is important because it is going to take a long time for the Earth to receive the right number of heavy elements, including radioactive material, and for life to develop and mature. Irregular galaxies are often destroyed by other star systems. Elliptical galaxies, which are much older, burn out too soon, producing a supernova.

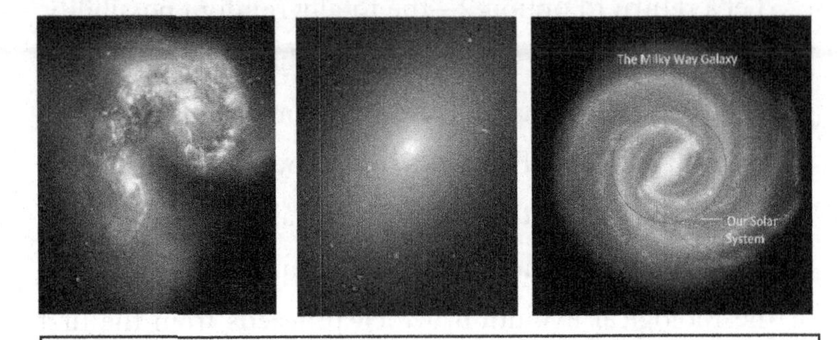

Figure 4.1 Irregular, Elliptical, and Spiral Galaxies. Note the circle drawn for the spiral galaxy. This is the so-called co-rotation distance for the spiral. Solar systems at this distance are very stable in their orbit.

By examining the composition of the Earth, scientists have noted, by comparison to the rest of the Milky Way Galaxy,[28] that the Earth has far more heavy elements than one would expect. Some unusual abundances of heavy elements include Fluorine (50 times more), Sodium (20 times

---

[4-28] Hugh Ross, *Improbable Planet.* (Grand Rapids, MI: Baker Books, 2016), 167.

more), Aluminum (40 times more), Potassium (90 times more), Titanium (65 times more), Nickel (20 times more), Copper (21 times more). Think of how much aluminum and copper is needed to meet our modern society needs. In addition, having heavy elements is a good thing as seen by examining your daily vitamin pill labels where one finds Zinc, Selenium, Copper, Manganese, Chromium, etc. We need these heavy elements, but we don't need too much. Too little manganese can result in serious diseases such as arthritis, diabetes, osteoporosis, etc. Too much manganese results in headaches, tremors, loss of appetite, muscle rigidity, leg cramps, and hallucinations. We live in a delicate balance.

So how did the Earth get these extra amounts of heavy elements? Heavy elements require extra energy to become fused with lighter elements. This occurs in Supernova, when a star with the right size and composition effectively blows up. According to one theory, our solar system moved or was formed closer to the center of the MWG. That was a much hotter, more radioactive place than the place where our solar system is now. This allowed our solar system to receive a more unusual number of heavy elements.

The center of the MWG would be a difficult place for man to live; fortunately, our solar system for some reason moved away from the center to a place between the arms of the MWG right at a critical distance where the solar system would neither move back to the center nor get flung out into space and also receive just the right amount of radiation. As a side benefit, this location allowed the future human observers to gaze out into the universe to see the grandeur of the universe and determine aspects of its origin. Being within a spiral arm would have made it too bright for us to see other galaxies. Being in just the right spot to get extra heavy elements and then to move to the ideal location for Earth and life to develop was indeed fortuitous.

The Water Habitable Zone

At this point, after 9 billion years since the Creation event, we find that the solar system is at an almost perfect location. Is that enough? No, it is not. The Earth itself must be an ideal location within the solar system. There are many hazards in the solar system, and the Earth must somehow be protected. The Earth obviously must be habitable. Water is a major component for life to exist, and water must exist in all three phases—gas, liquid, and solid. Thus, the Earth may be neither too close nor too far away from its source of heat—the sun (another Goldilocks requirement). Scientists have stated that if the Earth were only 1% closer to the sun, the Earth would have runaway evaporation of water. If 37% to 67% further from the sun (depends on the model used), runaway freezing of water would occur.

Even if the Earth were to be at the right distance, it must stay at the right distance in a stable orbit for life to develop, continuing to become more and more complex. The orbit itself must not only be stable, but it must also be nearly circular. Fortunately, the Earth's orbit around the sun is nearly circular. As it is, people now complain that the Earth gets too hot or too cold just because the Earth tilts toward the sun in the summer and away from the sun in the winter. Can you imagine the temperature swings if the Earth's orbit was more like an American football, as many of other planetary orbits are! Life as we know it could not exist.

And what about that little wobble we have? Another fortuitous development. Even with the wobble the Earth has, it stills has ice caps at the north and the south poles. We would be limited to life just around a central band above and below the equator. The wobble effectively evens out the temperature as much as possible as the Northern Hemisphere gets warmer in May to September and the Southern Hemisphere warms up in November to March. A few pages back, I mentioned that with the spin of the Earth, we can define the concept of a day. What if the Earth didn't spin? Our moon doesn't spin. We see the same side of the moon each and every day the moon is visible. With no rotation, one side of the Earth would be extremely hot, getting no reprieve from the day's heat while the other side of the Earth would be entrenched in a permanent cold freezing night.

Fortunately, the Earth rotates, but you may be surprised to know that the rotation is slowing down due to the gravitational effect of the sun and the moon. This has been going on for a long time. 4.5 billion years ago, it has been estimated, the Earth rotated once around in only 3 hours instead of the current 24 hours. Can you imagine the wind conditions in those days? Someday, if the Earth can avoid catastrophe, the Earth will rotate once every 48 hours. Twenty-four hours of sunshine, 24 hours of night. Think of all the rounds of golf one could play every day! On the other hand, the noon day siesta to escape the heat would have to be expanded. Oh, well, I definitely will not be around at that time, billions of years from now.

The question remains—will mankind still be here? That is pretty doubtful considering what the sun may be doing at that time. We will have global warming for sure. In fact, we have global warming now, not because, as some scientists believe, of an excess carbon footprint due to the consumption of fossil fuels, but because the sun is simply getting hotter and hotter. The sun will continue to get hotter no matter what we do. That is not to say that we aren't exacerbating the problem now. Fortunately, we live at an ideal time for man's survival. We are so lucky!

Other Habitable Zones

Water isn't the only concern. Over 200 requirements exist that science has discovered. Without any other of these requirements, man's existence is at best questionable. Here are a few examples (some have been partially discussed already):

The UV habitable zone – Low levels of Ultraviolet radiation are necessary for man's existence. As an example, UV is necessary for the production of Vitamin D and for DNA repair. High levels are deadly for man. Skin cancer is but one example. Only 3% of the 40+ billion planets in the MWG meet this criterion.

The Photosynthetic Zone – We obviously need photosynthesis as this is how carbon is extracted for plant growth, and O_2 (Oxygen gas) is released to breathe. For this, the proper light intensity, ambient temperatures, and seasonal stability are required.

The Ozone Zone – Just the right ozone level needs to be present to help maintain a proper Earth temperature. This requires a correct O_2 level, and not only the right UV level but also a stable UV level.

A Correct Rotation Rate – As discussed already, this will affect winds, but it also affects the reflectivity of clouds, which also affects the temperature of the Earth's surface.

Obliquity Zone – As discussed, a planet needs the wobble to increase the viability zone on the surface of a planet.

Tidal Zone – If the planet is too close to the sun or to some other solar system object (as the moon is to the Earth) there will result a so-called tidal locking (no rotation), resulting in half the planet being too cold, and the other half being too hot.

Astrosphere Zone – Stars produce a "wind" of particles that can deflect/absorb cosmic rays. Too close to another star or star clusters will be deadly, and too far is ineffective. We need those particles, both radioactive and non-radioactive.

Other Earth Requirements for Life

Special Water Property – Going back to the creation of the constants for electromagnetic forces, we see a special property exists for water. It expands and gets less dense as it freezes. If water were any denser, the ice would fall to the bottom and eventually the lake would be one huge solid ice cube, formed from the bottom up. As ice is lighter and so floats, the surface of water will become frozen solid. This insulates the water below, protecting life below the water's surface and keeping the water warm. Not only does this protect marine life, but it also helps keep the Earth warm.

Total Water on Earth – The Earth was once a water world, completely covered by water. The collision of the Earth with a Mars-size asteroid, plus other events, resulted in large amounts of water being sent off into space. Now, only 71% of the surface of the Earth is covered with water. This is useful as we have rivers, oceans, dry land, trees, farms, etc., which allow us to develop and transport our food systems.

Essential Magnetic Force – The Earth naturally, because of all the molten iron moving beneath its mantle, has a magnetic field. The magnetic field deflects away solar particles from the sun (a solar wind). In order for the magnetic field to be generated, the molten iron must be able to move around. Thus, we find the molten iron has just the perfect viscosity and temperature to do this. Scientists tell us that this only works if a planet's mass is within 1.0 to 1.4 times that of the Earth.

Habitable Longevity – It takes a long time for everything to develop. Solar system changes must happen within time frames allowing the Earth to adjust. The Sun used to be hotter at one time, thus the planet Mars may have been warm enough to have running water, but the sun got cooler, and water only exists in a frozen state on Mars (if at all). Life developed on the Earth during a perfect time. It won't last, but the time frame is long enough for life on earth to flourish. As it is, even solar flare activity is at a minimum at this time. If nothing else, this lets cell communication be relatively static free, so our children can text or play games on their phones all day long (not a good idea)!

Helpful Moon – Our moon is a wonderful thing to have. Poets, songwriters, lovers, even wolves have long spoken of the beauty of the moon. It is nice to have such a light in the sky for much of the night most days each month. Many other planets also have a moon or moons. Our moon is special, but almost didn't happen. In relation to the size of the earth, our moon is much larger than other moons. Why did this happen? Many scientists believe that the moon was created due to collision of a Mars-sized asteroid with an Earth smaller than it is now. This collision could have broken the

Earth apart if the asteroid was moving much faster and if the collision was more direct. As it turns out, there is a narrow range of angles and speeds which would allow the collision to be successfully achieved. As noted above, much of the Earth's water and a substantial amount of material was splattered into space, only to later be re-formed due to gravity as the moon. (Thus, because life was already existing in unicellular forms on Earth, it is likely that future studies of moon rocks will show the remnants of early Earth life.)

The Moon also creates tides which cleanse our beaches every day while also slowing down of the Earth's spin rate (together with the Sun's gravity) and stabilizing the Earth's wobble. All due to gravitational effects. (Don't forget how important establishing the Gravitational constant at Time=0 was.)

Helpful Planets – We are fortunate to have a large planet relatively close to us. Jupiter, which is more than 300 times more massive than Earth, is close enough to gravitationally attract many large asteroids away from the Earth. I personally would hate to see a large asteroid heading our way and am grateful to have Jupiter sweep away all the asteroids, comets, etc,. heading our way. I am sure the dinosaurs wish that 67 million years ago, Jupiter would have done its job. It just didn't work out too well for them.

Tectonic Plates – This is very important. The collision of the plates and the volcanic activity continually recycle the Earth's crust, providing new minerals and releasing gases.

This presence of a molten center and plates is due in part to all the radiation received in the Earth's early years. (Radiation from radioactive elements increases heat as part of their decay.) Venus and Mars by comparison have solid crusts and are not revitalized.

Past Bombardments – 3.9 billion years ago, the Earth had what is now known as the Late Heavy Bombardment by comets, asteroids, heavy elements, and other space materials. This resulted in 200-1000 tons of heavy elements and material PER SQUARE YARD being deposited across all the Earth. This raised the temperature of the Earth melting/vaporizing much of the surface ice/water. More than 10% of the earth's surface rocks were melted down to 1 km deep. Was this important? As the Earth's surface cooled, single-cell life flourished.

And On and On

The list of what is needed for man to thrive and develop civilizations keeps getting longer and longer as more and more research examines what is really necessary for intelligent life to exist. There are now over 400 different conditions which scientists have identified which need to be fine-tuned in our Galaxy-Sun-Earth-Moon system in order to support life. The likelihood of all the conditions occurring for a single planet has been calculated to be less than 1 in 10^{1032}.[29] We truly are the lucky ones. No other planet in the universe has been found to meet these criteria. So now we have both a universe whose constants are fine-tuned and a cosmic evolution that is fine-tuned with all its "Goldilocks" requirements—all to make life possible. It's not reasonable to think it could happen, yet life did begin to exist on this planet we call Earth. For those who want to go deeper into the physics of our Universe and all the fine-tuning involved (without having to struggle through all the mathematics involved), I recommend Lewis and Barnes' A *Fortunate Universe.* (See Books of Interest for details.)

Let me end this chapter with a quote attributed to one of the world's best-known atheistic scientists—Richard Dawkins.

> *"The world easily could have remained lifeless. It is an astonishing stroke of luck that we are here!"*

Imagine that, a scientist believing in LUCK!

In the next chapter, I will discuss many other quirks of existence that have been part of the story of the development

---

[29] Hugh Ross, *Why the Universe is the Way It Is (*App C: Baker Books, 2008).

of rational life. Once again, everything seems to be totally unlikely, but it happens.

Things to Think About

1. What new knowledge have you learned about the heavens and the Earth since you were a child? How much was this knowledge gained by you figuring it out?

2. This chapter noted some special property of water. Can you name any others?

3. Some people just seem to be lucky. Is being lucky really an attribute for some?

Chapter 5

If He Builds It, They Will Come!

In the last two chapters, I spoke about the preparation of our Universe and of the Earth so mankind would have a suitable place to live. Of course, preparation would be meaningless for us unless mankind came into existence. Once the Earth came into existence, how did life arise? Did life come into existence as fundamentalists believe in three days (Days 3, 5, and 6 in Genesis) 6000 to 10000 BCE depending on whom you talk to? Or did life come to arrive over a long period, approximately 3.9 billion years? Can two completely opposite theories both be true? When considering this issue, remember one has to view the FULLNESS OF REALITY as was discussed in chapter 2. Don't have a mindset of only science or only God. We are looking for the FULLNESS OF REALITY, the FULLNESS OF TRUTH. If you limit the possibility to only Science or only Scripture, that is, to only your pre-disposed point of view, you may end up, while still being a nice person, becoming willfully blind to the true and complete facts of life—facts, which are not explainable in your mindset.

Now, neither side can effectively refute the other, when it comes to recognizing God and the transcendent nature of man. If God is real, then by definition, He could create life and anything else He wants. Belief in God covers everything, but it sure would be nice to know how it was done. Science, on the other hand, continues to understand more and more about the life process, including what it means to be alive, what is necessary for life. It's a fascinating topic, but one with issues.

Science for its part has not been able to create life from inorganic materials. Many theories abound, but none has definitively been able to duplicate or even conceptually show to the satisfaction of all scientists how inorganic materials became organic materials. Like Physics in regard to what happened after the origin of the Universe critical event (or the Multiverse, for that matter), biology (so far) only knows what happened after the "origin of life" critical event. Both sciences are after the fact sciences. For the honest biologist, how life got started remains a mystery, a conjecture at best.

And, what should happen should biologists somehow demonstrate the development of a living cell out of inorganic materials (and not just some hand-waving argument)? The faithful will always be able to say that life can only occur because, in the beginning, God created the Universe and established the laws of physics and their associated constants, which I discussed in chapter 3, and oversaw the process such that the constants allowed inorganic materials to eventually form a living cell.

Remember from chapter 1 that all knowledge is based on how much we trust our sources of knowledge—whether it is knowledge from our senses, our experiments, our trust in someone, our experience, et cetera. The deeper our trust is in our sources, the more we are likely to say we know something. Eventually, we must step over the line and realize that we know something ourselves. For example, while I have never been to Russia, I, based on my trust in people, books, movies, pictures from space, etc., must say I know Russia exists, or else I would perhaps be considered psychotic. So, for now, let's not get worked up into a tizzy. I am not going to prove anything beyond a shadow of a doubt; at best, I'll just see what knowledge is out there. You will need to determine what you have really come to know.

Let's look at three possible scenarios:

1. From a God-driven point of view, one might say, "God created life in 3 different days (days 3, 5, and 6 in Genesis) but made His creation to appear that it is older than it seems."
2. From a God-driven point of view, one might say, "God, for whom '*one day is as a thousand years, and a thousand years as but one day*' (2 Peter 3:8), created life, but chose to have life develop over time. Let's see how it was done."

3. From an atheistic non-God-driven view, someone might say, "We don't really know how this happened. But we are sure it must be an unintended process. Let's continue the research."

I will keep an open mind but will only have a limited discussion of the first scenario. This may be a personal bias, or maybe I'm just willfully blind in this area. I just keeping asking myself, "Why would God do that?" For me, the fact that Jesus said He is "*the way, the truth, and the life*" (John 14:6) is good enough for me. There is just so much evidence which indicates the Universe, the Earth, and mankind are way older than 6000 years of age. Deceiving scientists just doesn't seem to be the thing God who is truth would do. Many, and I would say most, Jews, Christians, and Muslims are perfectly comfortable accepting that Genesis was written to reveal WHY God created and the ramifications of that act, and not to explain the specifics of HOW God created.

The second and the third options reflect the real issue for those who want to know the HOW and not the WHY of the Origin of Life. The answer sought is, "Is the cause of life an accident, or is it intentional?" Plus, as an add on, how was it caused? In a way, this is the same as the problem of the origin of the Universe. No one knows scientifically why the Universe started, much less why it started as it did. One can only study the evidence one sees and conjecture. What happens after life originated is a somewhat different question than that of Origin of the Universe in that there is a physical world which precedes life, whereas, prior to the Origin of the Universe (or the Multiverse), there was no physicality nor even time. Perhaps, as some believe, the physical world was destined (by whom???) to bring about life.

This question of how the spark of life arose is addressed scientifically as Chemical Evolution. What happened after life began, then, is Biological Evolution. This chapter will address Chemical Evolution—that moment of time when the first life began from inorganic materials.

Most scientists working in this area seem to agree that life has existed on Earth for at least 3.9 billion years. However, that does not mean that life originated then. Some say it could be even earlier. Scientists have discovered that the oldest rocks already had anaerobic unicellular life.[30] Furthermore, scientists know that life is even older than the oldest rocks because they can tell that the cells they have found came from other living cells. What? How can scientists know whether the first cell's "parent" was a rock or a living entity? It turns out that our carbon-based life forms have an interesting quirk. Living matter prefers to be preferentially comprised of the Carbon-12 isotope (6 protons and 6 neutrons) as compared to the Carbon-13 isotope (6 protons and 7 neutrons). And guess what, the cells from the earliest

---

[30] Niles Eldredge, *The Triumph of Evolution and the Failure of Creationism* (New York: W. H. Freeman, 2000), 35-36.

rocks also show a preference for Carbon-12 over Carbon-13, so these cells arose out of already existing life-bearing cells. Thus, one has to admit that we don't really know how old the first life is. Could life have existed on rocks that must have been around even earlier? Some say we may never know. This is due to an event called the Late Heavy Bombardment (LHB), which essentially melted the Earth's surface. Still, there are other scientists who say that life may have essentially existed at the formation of the Earth itself. But how could that have happened if life was created by a random process? That has to be awfully unlikely. We'll discuss that likelihood in a few paragraphs.

But wait a moment, everyone who understands evolution, knows that evolution is based on the concept that evolution is not a series of random events but must include "gradual, cumulative, step-by-step transformations from simpler things."[31] Everything must be gradual, cumulative. But that concept relates to Biological evolution and not to Chemical evolution. Chemical evolution moves from no life to new life. In a similar situation in life, a woman is either pregnant or not pregnant. There is a big step from having so-called prebiotic molecules swirling around, to having the same molecules combining with other molecules, including many proteins, complex sugars, nucleic acids, peptides, and

---

[31] Richard Dawkins, *The Blind Watchmaker* (London, UK: Penguin Books, 2016), 14.

at least RNA, and becoming encased in a cell membrane all at the same moment of time. That is the challenge that has yet to be duplicated. For now, let's take a step back and talk about the concept of life itself.

In recent years, Science has made great strides in describing what is life and what is necessary for the life processes to exist, and how to continue life. Some scientists believe there are 7 criteria for life. All living beings must:[32]

1. Be able to respond to stimuli
2. Grow over time
3. Produce offspring
4. Maintain a stable body temperature
5. Metabolize energy
6. Consist of one or more cells
7. Adapt to their environment

This is a fairly complex summary of the elements of a living cell. Obviously, a rock doesn't meet this criterion, even if it gets larger and larger due to additional deposits each year. Viruses are also arguably not alive because they require a host to replicate; that is, without a pre-existing living host viruses cannot replicate, and without the entity replicating, the set of molecules is just an anecdote in history. From the

---

[1] Laura Geggel, "Are Viruses Alive?" (February 25, 2017), at www.livescience.com/58018-are-viruses-alive.html.

first existence of a unicellular life, replication had to have been obviously necessary, as would have been the metabolizing of energy, and the other items on the list. How did all this happen?

Some believe that maybe life began in a primordial soup of prebiotic molecules. This soup includes many of the common amino acids, peptides, even RNA-like molecules, etc., which are part of everyday life. They may have been just floating around, until one day they combined into a cell. Is that realistic? Sounds perhaps feasible, but there are some issues. First of all, there is the problem of the stability of prebiotic molecules due to something called the oxygen-ultraviolet paradox. It has been found that the presence of oxygen severely stymies the development of prebiotic chemistry, so that makes pools of prebiotic molecules floating around waiting for the right moment to combine just not possible. However, if there were little or no oxygen present, there would be no ozone layer to protect the surface of the Earth from the Sun's ultraviolet radiation. Ultraviolet radiation would also be catastrophic to any prebiotic molecules.

There are many other issues with the thought that prebiotic molecules were randomly formed. For example, the idea that RNA and many other molecules just hung around until somehow a protein cell wall formed around them fails to recognize that, if formed in water—whether an ocean, a stream, or even a crack in the soil, etc.—the molecules would hardly stay together due to the agitation, motion, evaporation, etc., of the water. Besides, water is a solvent, meaning that it tends to dissolve material rather than cause molecules to remain together. Large molecules would not stick together long enough to be gathered into a cell.

To combat this thought of solvency of water and turbulence in the water, some scientists have proposed that the chemicals came together in cracks in a dry clay surface. These surfaces might have had mineral crystals that could have organized organic molecules, concentrating them into organic compounds. After a while, organic molecules could have taken over this job and began to organize themselves.

But realistically, how would the molecules organize themselves? It would take a really long time (perhaps a million years) for such a thing to happen, and wouldn't weather, floods, UV radiation, and even asteroids pounding the surface of the earth soon disturb any progress that was made.

Thus, for these and more reasons, the once common thought that life arose out of prebiotics falls apart. There would be simply no primordial soup from which living cells arose.[33] Plus, how could so many biologically necessary molecules happen to get together and then magically become surrounded by a cell wall or at least a cell membrane.

---

[33] Ross, *Improbable Planet,* 97.

Think about it. It would be pretty eerie to think of all the molecules necessary for a cell to metabolize, react to stimuli, and replicate, etc., to be floating, disconnected, in some pool of water or stuck to clay. Suddenly, a non-destructive lightning bolt or other process happens, and there is magically some kind of membrane or wall which forms around the molecules and voilà, a cell is formed!

Even more, besides the mechanics of such a thing happening, there is the chicken or the egg problem. The problem is that cell walls and most other essential components of a cell are composed of proteins, and to make proteins one needs DNA, or at least RNA. To make DNA or RNA requires proteins as part of the process. But, keeping an open mind, maybe it is possible.

Information in the Cell

All the problems of getting the right molecules there at the right time, however, are peanuts compared to the real issue perplexing scientists. The construction of RNA and DNA requires **information**, and information is not physical. Information is not random. Information implies knowledge. How does that come from random particles? So, how does the construction of DNA and RNA imply knowledge? So, bear with me. This is not a textbook on molecular biology. I am not going to say in a few paragraphs what people spend years to learn. Instead, I will by example show you how the whole RNA and DNA thing works and why information is so important.

First, it is important to note that information must be passed on from one generation to another. In this case, the information we are talking about are pretty basic things—how to reproduce, how to manufacture goods, how to repair broken parts, how to regulate energy, how to replenish energy, etc. In one point of human history, information was passed through hieroglyphics. This required the interpretation of images. This obviously will not work on the cellular level. Later man developed ways of writing things down using a series of symbols and letters which together create words. From generation to generation, man's information has been passed down through a written language.

Believe or not, no—just believe it—the cell also contains a written document using both letters and words which are found in both RNA and DNA molecules. No one who knows anything about RNA and DNA denies this. Let's see how that can be.

RNA is a single stranded molecule and DNA is a double stranded molecule. For our purposes, just think of a strand as a simple string. True, it is a very thin string, about 1/40000 the thickness of a human hair. DNA has two strings side-by-side, twisted about each other like a thread with electrical bonds which keep them from untwisting. As one would expect, a single stranded RNA molecule is far more fragile than a DNA molecule.

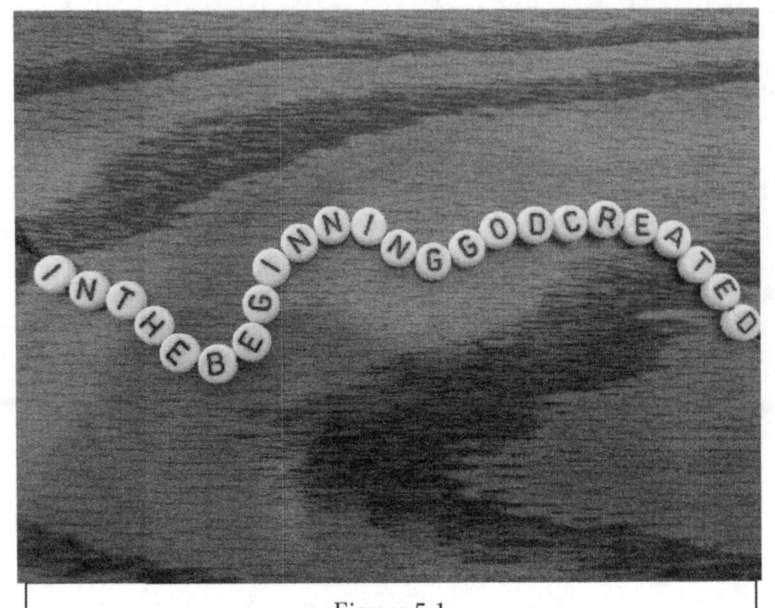

Figure 5.1
Information contained by placing letters onto a string.

Now for the information part. In our English language, we use 26 letters, plus spaces and other punctuation marks. We combine the letters to make words and separate them from each other with some form of punctuation mark. We continue adding words, and before long we have sentences, then paragraphs, and finally a document. Maybe the document that is written will tell us how to make the components of a car, put the parts together, add gasoline, and on and on. Instead of writing the letter on a piece of paper or in a PDF document, we take various beads with letters or punctuation marks on them and slide them on a string. If the string is really long, we can form whole paragraphs. Figure 5.1 shows how one could put the whole Bible on a string if one chose to. (Note: There are no spaces shown, just as the original Hebrew scripture and Greek New Testament had no spaces or other punctuation. Similarly, our RNA and DNA have no spaces between words.) Copying the document just means that one has to somehow duplicate the string.

In RNA and DNA, there is also a language. Former President Bill Clinton called it "the language in which God created life!"[34] Only for DNA and RNA instead of 26+ letters/punctuation marks there are only 4 letters. In our human language, there can be as many letters in a word as needed. One then puts as many words together as necessary (and sometimes far too many words—my apologies!) in order to create a meaningful sentence or paragraph.

Necessarily, if the letters or words are in a random sequence, then the paragraph will be meaningless. Furthermore, in "God's language," there are only 3 letters per word (*antidisestablishmentarianism* would not be allowed), and only 20 different words allowed. That's a small vocabulary! In biology, instead of beads with letters on them, the "letters" used are instead the nucleotide bases **A**denine, **C**ystine, **G**uanine, and **T**hymine for DNA. For RNA, the "letters" are **A**denine, **C**ystine, **G**uanine, and **U**racil. (What are nucleotide bases? Don't worry about it. The actual chemical make-

---

[34] Francis S. Collins, *The Language of God: a Scientist Presents Evidence for Belief* (Free Press, 2007).

up is not important for what we are to discuss. Wikipedia can explain it all if you wish to go there.) Just remember the letters **A**, **C**, **G**, **T**, and **U**. The 3 letter combinations—the words—biologists call these words **codons**. For instance, the 3 letters **GCU** would be a codon, and that codon corresponds to the amino acid Alanine. See Figure 5.2 below for an example.

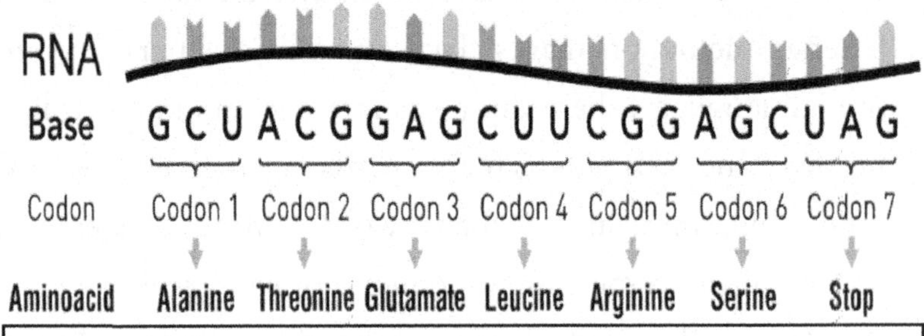

Figure 5.2 Schematic of a single-stranded RNA molecule illustrating a series of three-base codons. Each three-nucleotide codon corresponds to an amino acid when translated to protein.

Now compare the 20 words allowed for RNA/DNA to the average number of words known by a typical 60-year-old person – 48,000[35]. The number of letters and words allowed for DNA/RNA would seem to be pretty limited. However, nature has overcome this limitation of a vocabulary of 20

---

[35] Wikipedia, "Vocabulary," available online at aten.wikipedia.org/wiki/Vocabulary as of March 28,2020.

three-letter words by allowing "sentences and paragraphs" to be as long as necessary. This might mean there may be needed **hundreds of thousands** or more 3 letter words stranded next to each other in the RNA/DNA molecule in order to convey the instruction intended. This turns out to be very powerful, and dare I say, also very efficient method. However, personally I prefer much shorter sentences when I communicate.

But why should there be only 3 letters and 20 words? Proteins, we know, are essential for all Earthly life. Each protein molecule is essentially a chain of amino acids—one amino acid after another in a very specific sequence. Now, one actually doesn't need to know what an amino acid is, after all early man, who was perhaps as intelligent but not as knowledgeable, got by for over 100,000 years not knowing what an amino acid is. Suffice it to say you have probably heard of many amino acids—lysine, leucine, tryptophan, etc. I am sure you have heard of tryptophan because every year we hear on Thanksgiving Day that tryptophan, found in turkey, may make you sleepy.

Now here it is where everything all comes together. Remember those codons, those words written in sequence on the "string" of RNA/DNA? Each codon matches up to one, and only one, particular amino acid. How many different amino acids are used to make proteins? You have already guessed it—20! The codons on the string are used to create the chain of potentially 20 different amino acids which make up a protein molecule. How this happens is outside the scope of this book; but it works! If one could lay an RNA/DNA strand next to the final protein molecule (scaled down, as proteins are much larger), you would find a perfect match. Voilà! If the sequence of codons is changed, one will have a different protein or no protein as the protein may fall about otherwise.

Furthermore, the order of amino acids in each protein will result in a different specific shape. Some will be straight, like a hair protein. Others will take shape in what is called a quaternary structure, such as hemoglobin. If you have the wrong shape, the protein will not function as intended. The final shape is very dependent on the sequence of amino acids.

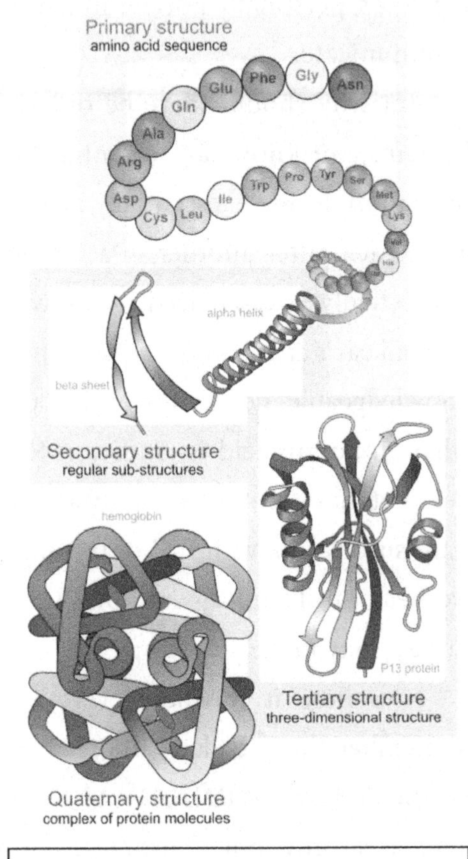

Figure 5.3 Basic Protein Structures

So far, I have shown how a series of codons can make a protein. DNA with the help of RNA controls everything. In speaking of the strands, I mentioned that the use of the 20 words (comprised of 3 letters each) would result in very long sentences to instruct what needs to be done. Man's DNA has **3.2 billion** letters (base pairs) strung along its strand. Certain sections are used to define specific aspects of an individual; others give instructions to perform specific tasks at different times as needed. Figure 5.4 shows my version of a string with beads giving a command. It is easy for us to read my instruction using our eyes, but it is not quite as clear how a command is properly interpreted, say, to repair a section of the cell wall in a real live cell.

Figure 5.4 Beads on a string giving a command.

One or more sections of DNA from different parts of the various strands determines whether the eyes are blue, hazel, brown, etc. The same is true of other aspects such as the color and type of hair, skin, even of personality. Other sections of DNA will have to be activated in the early embryo in order to develop different types of tissue, such as nerve, muscle, kidney, eyes, etc. Much later, whole sections of DNA must be turned on to start puberty or to end fertility. It is as if the various sections are *hyperlinked*, as one now finds in so many documents on the internet. The various cell molecular machines that do all the necessary cellular work just know how to hop from one place on the DNA strand to another. DNA has a big job to do. Each of the activities will be controlled by one or more series of codons within the DNA strand. The series of codons which control such activities are termed genes. Thus, we can speak of dominant and recessive traits found in genes. Breakdown in the DNA sequences due to incorrect copying or chemical destruction/replacement often results in deficient activities, resulting in genetic diseases. As you can see, DNA and its helper RNA are physically important to us all.

I have spent a fair amount of space describing the strands of RNA and DNA, and why they are like they are. I assure you, by now, I am almost out of breath describing the process. It is such a clever system, but hardly a simple system.

The Origin of Life is such a mystery. It needs to be contemplated. While contemplating, let's not forget what we have already discussed in chapter 3, that if the Electromagnetic force established at the moment of the Origin of the Universe were different at all, none of this would have happened. Ah, but there is still more to discuss. In the next chapter, we will look at other issues surrounding the Origin of Life.

Things to Think About

1. Why is the Origin of Life not an evolutionary process?

2. The concept of the requirement for information for the existence of life is a game changer. Why is that?

3. How are computers conceptually similar/different from DNA?

Chapter 6
Life Can't Wait Forever

In 1966, a pop rock band by the name of *The Outsiders* had a hit song by the title of *Time Won't Let Me.* I have mutated the lyrics to better suit this chapter:

> "*Life* can't wait forever
> Even though you want *it* to
> *Life* can't wait forever
> To know if *Evolution is* true
> Time won't let *it*
> Time won't let *it*
> Time won't let *it-it-it-it.*"

In chapter 5, I showed how unlikely it was, from a random point of view, for life to come into existence in a cell. I didn't address the cumulative approach as there was no previous life to improve upon. However, some have argued that there could have been a cumulative effect for RNA. RNA could have reproduced itself and, over an unknown period, produced a new, more sophisticated RNA molecule with all the necessary information to make itself and other proteins, including a cell membrane to encase everything. It is also, as I have stated previously, unlikely that previous unlinked molecules happened to develop and all come together at the right time to initiate a reproducible, functional cell. Well, you might say, that is just your opinion that it couldn't have happened. You might employ the Argument from Personal Incredulity, as Richard Dawkins often does,[36] when trying to argue with those who believe that the odds of Neo-Darwinism, based on common sense, is so low as to be effectively zero. Technically, the Argument from Personal Incredulity is a fallacy in informal logic. Arguments from Incredulity can take the form: [37]

1. I cannot imagine how F could be true; therefore, F must be false.
2. I cannot imagine how F could be false; therefore, F must be true.

(This same argument will be argued against Richard Dawkins when we discuss his thoughts on miracles.)

So, the question is, "Besides common sense, is there any evidence that will show the random/cumulative approach

---

[36] Richard Dawkins, *The Blind Watchmaker*, 38.

[37] Wikipedia, "Argument from incredulity," at en.wikipedia.org/wiki/Argument_from_incredulity (04/19/2020)

can't work when applied to the Origin of Life?" The answer is YES, because *Life can't wait forever. Time won't let it.* We must examine the time factor. To be fair, the research on the time it would take for the evolutionary processes to occur under random mutation processes and cumulative effects wasn't calculated until recently. The argument had always been, "Given enough time, nature can do anything." Well, the Earth is still a young puppy considering the amount of time necessary to even begin the Origin of Life question.

To explain this, let's go back to the question of information. I assume you understand by now why information is such an important content of every RNA and DNA strand. In the Origin of Life issue, many evolutionary biologists and their cohorts have currently suggested that the earliest cell arose from the activity of a single-stranded RNA molecule, randomly composed. As there was no previous life, one cannot say that there was a benefit for one sequence of codons randomly assembled to be better than another. If eventually another codon sequence was added to the RNA strand, that would be wonderful. And, so the RNA would have to keep replicating itself, gradually adding codon sequences which could generate more and more proteins. Eventually, the RNA sequences would be so long that over 300 proteins including enzymes would be generated, plus proteins to make a wall around the RNA and the other proteins. Nothing to it! The problem is that this process would take an incredible amount of time.

RNA is not nearly as long a strand as DNA. In fact, most RNAs are no more than a few thousand nucleotide bases long.[38] Let's use a simple case: let's assume, somehow, that the RNA making the first cell has a sequence of only 250 nucleotide bases. Since only 4 nucleotides are used in constructing RNA, the probability of having the "letters" in just the right order mathematically is 1 in 4^{250} or, equivalently, on the order of 1 in 10^{150}. That is more than 1 in a trillion trillion trillion trillion trillion trillion trillion trillion trillion trillion trillion trillion—not very good odds. This, like the probabilities seen in chapter 2 in discussion of the origin of the Universe, is an incredibly small possibility, and the odds for success could appropriately be called 1 in a godzillion. Now think of that process happening for many molecules in a sustained cell. This is totally unrealistic to magically/luckily/randomly happen in the time frame that life is known to have existed on Earth. Time frame is an extremely important issue. The Universe, after all, is only 13.8 billion years old—still a young lad as far as its expected existence goes.

To see what I mean concerning the importance of the time factor, let's look at how including a time frame for the construction/destruction of a nucleotide base chain affects

---

[38] B. Alberts, A. Johnson, J. Lewis, et al. "From DNA to RNA," *Molecular Biology of the Cell.* 4th edition, (2002), at www.ncbi.nlm.nih.gov/books/NBK26887/

the rationality of random acts creating the first cell. From science, we know that life has existed on Earth for at least 3.9 billion years. On the other hand, the Earth is considered to be 4.5 billion years old, so only the first 0.6 billion years (600 million years) at most was available to try to create that first life.

As mentioned, most efforts to explain the Origin of Life have tried to show how various life molecules, such as amino acids, fats, lipids, and the nucleobases from which DNA and RNA are built, occurred in just a random process. This is totally unrealistic. Six hundred million years may seem like a long time, but when you consider the low possibility and the amount of time needed to create randomly just one 250 nucleotide RNA molecule, you will see, six hundred million years is just an instant of time. Life on Earth essentially arose on a cosmological time scale in a "flicker of an eyelash." So, let's see how much time it takes to randomly construct just one 250 nucleotide molecule and compare that time to the 600 million years science states life had from the formation of the Earth to the first living cells.

Consider a generous time frame of 1 second to assemble and then if the sequence is not perfect, disassemble a 250-nucleotide base strand of RNA. So, if 10^{150} assemblies are needed (as determined on the last page), and each assembly takes a second, that would be 10^{150} seconds, or about 3.2 x 10^{142} years. That's a lot of years needed to probabilistically create just one 250 nucleotide RNA molecule. We also know that probabilities and certainties are not the same. One could require that in order to win a $1,000,000, a coin must be flipped eight times, and each coin must come up heads. The odds are 1 in 256. It could happen the first time, or it could happen on the millionth time. Probabilities are just probabilities. But, of course, even if nature were so lucky, one molecule does not a cell make! Multiple molecules must be correctly made. Besides, consider how young the Universe is. Science states the Universe itself is only 13.8 billion years (13.8 x 10^9 years) old. There is a huge difference between 10^{142} years and 10^9 years. In other words, it ain't going to happen just considering random events. For life to occur randomly, there must be other processes or interventions involved. However, as of yet, there are no alternate scientific theories which do not also require random events. Maybe there will be one tomorrow. Maybe not!

Now, do the Evolutionary Biologists acknowledge the improbability of such an event happening? Again, going to Richard Dawkins for insight. Dawkins speaks of the apportionment of LUCK! Everybody should have some luck. If only life could have the initial luck to get started, he believes, evolution would take over from there. Life just needed to win the lottery against 1 in 10^{150} odds. It seems rather odd for a scientist who only believes in particle-ism to be reaching out for a non-physical concept such as LUCK! In his book *The Blind Watchmaker*, Dawkins chooses a 1 in a hundred billion billion billion as the upper limit (10^{29}) to be the point at which LUCK runs out.[39] As noted, we are way past that. "Way past that," is not even a fair comparison. You would have to add 1 trillion trillion trillion trillion trillion trillion trillion trillion trillion trillion to that. Besides Dawkins just made that number up! He admits that "Cumulative selection cannot work unless there is some minimal machinery of replication and replicator power, and the only machinery of replication that we know seems too complicated to have come into existence by means of anything less than many generations of cumulative selection!"[40] In other words, life needs a lot more LUCK than possible to get started.

And it doesn't stop there. The likelihood that life will begin and evolve to the point that we are today is so unlikely that Dawkins, himself, states: "If we don't use up all our ration of LUCK in our theory of the Origin of Life, we have some leftover to spend on our theories of subsequent evolution, after cumulative selection has got going."[41] So now LUCK not only exists, but apparently it is a ration-able quantity! It is not hard to see why when discussing evolution, evolutionary biologists often use the words, may, might, probably, possible, could have, my personal feeling, my hunch, just happens, et cetera, et cetera, et cetera.

---

[39] Dawkins, *The Blind Watchmaker,* 144.

[40] Dawkins, *The Blind Watchmaker,* 141.

[41] Dawkins, *The Blind Watchmaker,* 146

Other Issues — One Cell Line or Many?

As I have stated, many believe that life began shortly before or simultaneously with the oldest sedimentary rocks that came into existence 3.9 billion years ago. This means that life may have begun shortly after what is known as the Late Heavy Bombardment (LHB). In this event, the Earth plus Mercury, Venus, the Moon, and Mars were bombarded by so many asteroids and comets that an average of 200 tons of material PER SQUARE YARD was distributed over the entire Earth's surface.[42] Accompanying this physical destruction of the surface of the Earth was a massive increase in the Earth's temperature. Water evaporated off the face of the Earth (to be replaced later by other water containing comets and asteroids). If there was any RNA lying around, it would have been destroyed unless some cells had managed to find shelter in the crevices of Earth. If anything, it seems that the LHB served to essentially sterilize the Earth, providing a clean slate for life to begin about 10 million years after the end of the LHB.[43] Again, while 10 million years seems like a

---

[42] Ronny Schoenberg et al., "Tungsten isotope evidence from ~3.8 G year munavar more focused sentiments for early meteorite bombardment of the earth" *Nature* 418 (July 2002): 403; Ariel D. Anbar et al, "Extraterrestrial Iridium, Sediment Accumulation and the Habitability of the Early Earth's surface," *Journal of Geophysical Research: Planets* 106 (February 2001): 3219-36.

[43] Ross, *Improbable Planet,* 102.

long time to us, it is far too short of a time for random processes to result in the generation of life. Many biologists believe that life originated with one cell, but it is not clear whether that was the case. Perhaps it was one cell, or perhaps, there were multiple cells at multiple sites which experienced these life beginning events. No one can say if different cells didn't simultaneously develop. I will address this in more detail in chapter 7.

And, as an FYI, the LHB, besides sterilizing the Earth's surface, also included some last-minute fine-tuning of the Earth's composition. This "last minute" bombardment increased the Earth's core with extra sulfur, oxygen, iron, uranium, and thorium. Later, these elements would be useful materials for the future man's needs. How nice!

Instant Life - Panspermia

Some people believe that life was brought to Earth just after the bombardment—the concept of Panspermia, that the Origin of Life is alien. How foreign is that? Could that be true? Well, maybe it is. We can't definitively and scientifically say it is not. Maybe some alien beings sent out anaerobic bacteria mixed in with comets and asteroids to seed the Earth with life. The reasons an alien life force would do this can only be imagined. Of course, by sending out unicellular life, it would take almost 4 billion years for that life to develop to a point so that they could visit us someday—perhaps to make us slaves to them, or just to drop in for a friendly game of Bridge. Maybe their civilization was dying, and they wanted the Universe to have life somewhere—anywhere, including the Earth, so they sent out anaerobic bacteria in a capsule hoping it would survive the rigors of space and time, and perhaps evolve into higher animals. But that just doesn't make a lot of sense. First of all, I can't imagine civilizations thinking that far in advance, but that is not a very open approach—let's avoid the Argument of Incredulity! What advantage would that be to them? Besides, where did their life come from? We have already shown that the random component of making the necessary molecules would require a tremendously longer time than that of our Universe's existence, so how alien life would have developed anywhere else in the universe is as much of a question as how life started here.

Could life on Earth have started that way? Some say, maybe. Could God have been the means to bring life to Earth and any other planet of His choosing if He wanted? The faithful say, "for sure." Scientific materialists shudder at the thought.

Richard Dawkins unofficially spoke out for many scientists when he was interviewed concerning the Origin of Life in the documentary *Expelled:No Room for Intelligence* by Ben

Stein—a noted American writer, lawyer, actor, comedian, and commentator on political and economic issues:[44]

Ben Stein: *How did it start?*

Richard Dawkins: *Nobody knows how it got started. We know the kind of event it must have been. We know the sort of event that must have happened for the Origin of Life.*

Ben Stein: *And what was that?*

Richard Dawkins: *It was the origin of the first self-replicating molecule.*

Ben Stein: *Right, and how did that happen?*

Richard Dawkins: *I've told you; we don't know.*

Ben Stein: *So, you have no idea how it started.*

Richard Dawkins: *No, no. Nor has anyone.*

So, no one knows. Maybe science will solve the "How Life Began" question someday (but never the Why). Needless to say, the faithful already have. They call it a miracle. Beginning in chapter 8, I will show why people believe there is more to life than just being made of particles. For now, let's continue looking in chapter 7 at Biological Evolution and the development of life.

---

[44] Ben, Stein. *Expelled: No Room for Intelligence*, Documentary film (Premise Media Corporation Rampant Films, April 2008).

Things to Think About

1. Why do you think the time factor concern has only been addressed in the last 50 years or so?

2. What are your thoughts on Panspermia?

3. When asked how life began, Richard Dawkins indicated that no one has any idea. Is that true?

Chapter 7

You Say You Want an Evolution?

One thing I have learned is that when multiple, multiple people keep saying the same thing, there is a reason behind their beliefs. Let's take the Flat Earth Society. To me, it is obvious that the Earth is not flat, for I believe and trust the history books which say Magellan in 1519 led an expedition which sailed around the world (even though Magellan did not survive the journey). I also know that the top of the mast on a ship is what one first sees when one ship approaches another ship on the ocean. That is exactly what one would see if the Earth is round. I trust my eyes and my mind, so I can believe the Earth is round. I also believe and trust those who have taken pictures of the Earth from outer space, even though I personally have never been any higher than an airplane will take me. I trust that NASA and others are not trying to fool me. I trust them so much that I can say I KNOW the Earth is round. Knowledge requires Trust (I think I have mentioned that a few times.) Yet, I also appreciate that on a limited basis, the Earth does appear flat. The Bonneville Salt Flats in Utah certainly seem pretty flat, but as I stated in chapter 1, one must seek the FULLNESS OF REALITY. One can certainly come to the conclusion that the *Bonneville* Salt Flats are fairly flat on a limited scale, but in the bigger picture, most of the rest of us accept that the Earth is round.

As far as I can tell, nearly all biologists believe in a Biological Evolution which has occurred in which many, many changes were made in a slow, random, cumulative basis. The key word here is cumulative. Evolutionary biologists, such as Richard Dawkins, have stated that the evolution of life could not have occurred in a truly random process. He once stated:

> *"You may throw cells together at random, over and over again for a billion years, and not once will you get a conglomeration that flies or swims or burrows or runs, or does anything, even badly, that could remotely be construed as working to keep itself alive."*[45]

Darwin's slow random process which includes the survival of the fittest, however, does not fit all the data. It is perhaps a good start, but it's not enough. It is not that I am knocking Darwin. He was working with the information of his day. His theories have been superseded by what is today called Neo-Darwinism. Neo-Darwinism adds into Darwinism genetic variations due in part to mutations along with Mendelian

---

[45] Dawkins, *The Blind Watchmaker*, 9.

inheritance due to Mendel's laws of genetics. The evolutionary changes in life accumulated through slow, random steps. However, there is quite a bit of evidence which shows that evolution doesn't always proceed as a slow, random process. Often changes have occurred at a very rapid pace. Does that make Darwin wrong? No, the definition of "slow" simply needs to be redefined. A slow, random process doesn't mean that life changed in micrometers with each generation; rather, it means that changes were made in small discrete steps through random mutations. The changes that were beneficial became augmented. The changes that were equivocal may also have become the standard if for some reason down the road that adaption has a benefit. All these changes take time. If life conditions remain constant, such as competition for food, clothing, or changes in the enemies around

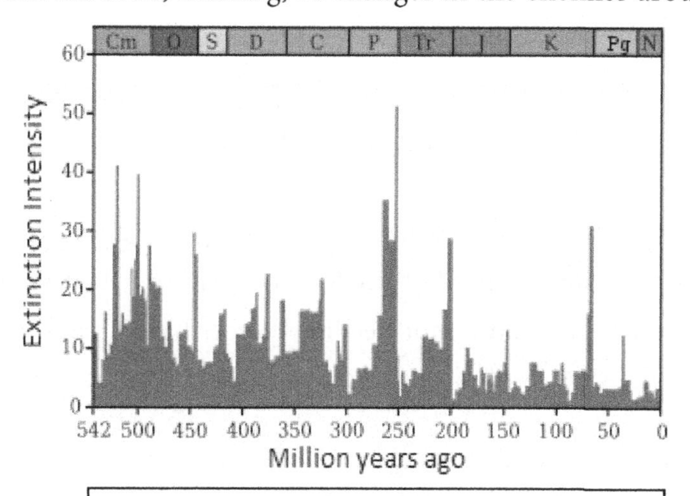

Figure 7.1 Catastrophic Events which have occurred since 542 mya (million years ago)

us, the slow speed may seem to hurry up as the outside world influences the necessity to grow.

Not surprisingly, science has found that a pattern of disaster followed by a change in life's characteristics has been repeated multiple times in our Earth's history. Figure 6.1 shows a series of catastrophic events nearly every 27 million years which has affected life here on Earth.[46] As there have been many catastrophic events involving narrow time frames, but which resulted in really good results as far as life is concerned, I will not be able to discuss them all. Furthermore, the intent here is not to detail all the geological events, but rather to open your mind to the possibility that these events resulted in positive benefits for the next biological populations (plant or animal), and perhaps they may have had a purpose. As such, I will only discuss a few such events.

The Neoproterozoic Oxygenation Event (NOE) and the First Avalon Explosion (~575 mya)

For the period 3.9 bya (billion years ago) to about 1 bya, life primarily consisted of single cell organisms. At first, the cells were anaerobic (not needing oxygen), then later with the gradual rise of photosynthesis there also arose aerobic

---

[46] Adrian L. Melott, and Richard K. Bambach, "Nemesis Reconsidered," *Monthly Notices of the Royal Astronomical Society Letters* 407 (September 2010).

cells which made use of oxygen. The cell structure gradually changed from being strictly prokaryotes (having no membrane surrounding the nucleus in the cell) to later also being eukaryotes (having a membrane surrounding the nucleus in the cell). Similarly, the Earth's atmosphere was initially primarily methane based—yes, the same gas that comes out of animals as flatulence and contributes to global warming, but this was also to change.

The first major explosion of life forms following the initial origin of unicellular bacterial life forms occurred after a series of glaciation events during the Cryogenian Era from 750 to 580 mya (million years ago). At that time, the Earth oscillated between being covered ice and warm interglacial events. The cooling of the Earth was due in part to the movement and breakup of supercontinents because of shifts in our tectonic plates which resulted in volcanoes, and the subsequent formation of huge basalt landmasses from the flow of magma.[47] The basalt landmasses reduced the heat as time went on, as the reaction of the basalt with water removed large amounts of carbon dioxide (CO_2) from the atmosphere, which lowered the capacity of the atmosphere to hold heat. As a result, temperatures plummeted. (Note: Global warming, which is occurring now, involves adding

---

[47] Y. Godderis et al., "The Sturtian 'Snowball' Glaciation: Fire and Ice," *Earth and Planetary Science Letters* 211 (June 2003): 1-12.

CH_4 and CO_2 to the atmosphere which increases the ability of the atmosphere to hold heat.) The warm inter-glacial events which followed occurred due to the spewing of CO_2 from renewed volcanic activity which followed, and with it the increase in the capacity to hold heat. Back and forth it went. By the end of the third major glaciation event—the Gaskiers Glaciation 579 mya, the atmosphere had changed from a methane (CH_4) dominated atmosphere to a carbon dioxide (CO_2) based atmosphere. This was accompanied by a warmer sun (the sun continues to get hotter every year due to hydrogen fusing into helium as the sun burns) and a corresponding increase in the sulfate-reducing bacteria. In a chain reaction, this increase in temperature and carbon dioxide allowed photosynthetic plants (which thrive on carbon dioxide) to flourish both on land and in the sea. The photosynthetic land organisms stimulated clay mineral and with that came phosphorus and other nutrients flowing to the sea due to erosion on the land. The addition of phosphorus to the ocean multiplied the photosynthetic organism. This resulted in a global oxygenation increase—now called the Neoproterozoic Oxygenation Event (NOE). With the large addition of oxygen, suddenly, in what is called the First Avalon Explosion, large, architecturally complex organisms developed.[48] Many were thin and sponge-like with large surface area to volume ratios (so-called Ediacaran). They lived

---

[48] Ross, *Improbable Planet*, 172-173.

primarily on the bottom on the sea floor. Other body shapes (some two meters in length) included fronds, discs, and segmentation. Whether one could call the Ediacarans plants or animals still remains unclear. What's more, there began symbiotic relationships between algae and bacteria and the various Ediacaran species—the algae and bacteria supplying the Ediacarans with photosynthetic products.

The Cambrian Explosion

Figure 7.2 Ediacaran fossil

Disasters continued to happen over time. Suddenly around 544 and 543 mya, a mass extinction event occurred, and nearly every Ediacaran species became extinct. Many attribute the mass extinction to the sudden release of large quantities of methane into the atmosphere caused by a shifting of the Earth's tectonic plates. This was followed by the most incredible change in life, which occurred during the famous Cambrian Explosion about 542 mya. At that time, 50% to 80% of all animal body types suddenly developed. Thus, for the first time, we find both vertebrate and invertebrate animals appearing, some with exoskeletons and internal skeletons which may be:

- rigid or flexible
- rod-based, plate-based, or some combination of the two
- made of calcium carbonate, calcium phosphate, chitin, or silica
- fixed, molting, or remodeling

This sudden change occurred because of a dramatic change in the seawater chemistry resulting in land erosions.

We suddenly found eyes with reflectors, lenses, and a cornea, which may be inset into the skull or stuck out on stalk-like structures. Compound eyes, like those of bees, appeared. The multitude number of eyes required complex image handling from each eye that then had to be interpreted in order to become one image in the brain.

We also saw some animals swimming in the open sea and others that simply inhabited the bottom of the seas. Predator-prey relationships arose almost instantly at the onset of the explosion. Of this period, it has been said by Jeffrey Levington, marine ecologist and evolutionary biologist: "No single environmental or biological explanation for the Cambrian Explosion satisfactorily explains the sudden appearance of much of the diversity of bilaterian animal life."

By Richard Dawkins, evolutionary biologist: "The Cambrian strata of rocks, vintage about 600 million years, are the oldest ones in which we find most of the invertebrate groups. And we find many of them already in an advanced state of evolution, the very first time they appear. It is as though they were planted there, without any evolutionary history."

By Thomas Cavalier-Smith, evolutionary biologist: "Evolution is not evenly paced and there are no real molecular clocks."

Permian-Triassic Extinction (252 mya)

About 252 mya, a cataclysmic extinction occurred. Many believe the source of this extinction was massive volcanic eruptions in Siberia which spewed out 1-4 million cubic kilometers (240,000 cubic miles-959,000 cubic miles) of lava. This would be enough lava to pave the entire surface of the Earth to a thickness of between 2 and 8 meters! Others believe that asteroids or comets bombarded the Earth at that time, or perhaps there was exponential explosion of anerobic methanogenic single cell organisms due to the addition of volcanic nickel to the water. The scientists still do not know. For whatever the reason, as a result, up to 96% of marine species and 70% of terrestrial vertebrate species became extinct.[49] In addition the water temperature rose 14° F to about 84° F for 5 million years.

While the exact cause of the extinctions is not known, what is known is that suddenly there were dinosaurs and mammals. Seed plants became dominant. Suddenly, there were herbivores, carnivores, parasites, and detritivores (fed on dead plants/animals/waste products) synergistically living together in beneficial ecologic harmony.

Cretaceous–Tertiary (K–T) extinction (66mya)

The last extinction I will discuss is my favorite one—most known for the extinction of the dinosaurs. This extinction is thought to have occurred due to the impact of a large comet or asteroid about 6 to 9 miles wide in the Gulf of Mexico, Yucatan Peninsula. Three fourths of the plant and animal species on Earth became extinct through what is called an impact winter. So much debris was thrown into the atmosphere by the impact that photosynthesis became limited. Plants need photosynthesis to live, and animals need the oxygen released by photosynthesis in order to breathe. All ectothermic animals (who rely on the environment to maintain their temperature) larger than 55 pounds, with the

---

[49] Wikipedia, "Permian-Triassic Extinction Event," (March 17, 2020), at https://en.wikipedia.org/wiki/Permian–Triassic_extinction_event.

exception of leatherback sea turtles and crocodiles, did not survive. Other terrestrial organisms, including some mammals, pterosaurs, birds, lizards, insects, and plants, in addition to those in the oceans, plesiosaurs and mosasaurs and devastated teleost fish, sharks, and mollusks were killed off or became seriously depleted. In the wake of the mass extinction, however, many groups underwent remarkable adaptive radiation—which is the sudden and prolific divergence into new forms and species within the disrupted and emptied ecological niches.

Fortunately for us, the dinosaurs were removed from the environment and many new forms of life further developed. How fortunate are we? Only 13% of the surface of the Earth is land. If the asteroid had arrived a little earlier or later, the rotation of the Earth would have resulted in a big splash, and no great cloud of dust to block out the sun, but the dust came and the dinosaurs didn't survive. Mammals, in particular, took full advantage of the disappearance of dinosaurs and diversified in the Paleogene period which followed—evolving new forms such as horses, whales, bats, and primates. Birds, fish, and perhaps lizards also participated in the diversification.

So, were these catastrophic events followed by major advances in the evolution of animals just coincidental? The appearance of new species was too quick to have occurred strictly by random evolution. Does this mean I don't believe in evolution? There is very strong evidence that evolution is a real process. To investigate this, we must look at the recent advances in molecular biology, particularly at DNA across all species.

Ways of Speeding Up the Process

It would seem that in certain events, particularly the Cambrian Explosion, evolution, if true, must have run amok. Suddenly, as noted earlier in this chapter, major, major changes appeared faster than standard expected mutation rates would suggest. Richard Dawkins, while not speaking of the Cambrian Explosion, hit upon a possible explanation. Dawkins indirectly addressed this when he resurrected the old question about a monkey randomly typing for as long as needed, eventually composing all of Shakespeare's works.[50]

Dawkins, himself, is quite willing to admit that if one considered only chance, there is no way monkeys could randomly type any Shakespearian phrase of reasonable length. What is needed to guide the monkeys (or a computer-generated program) is cumulative selection. In fact, for evolutionary biologists, that is what evolution is all about—random events modified by cumulative selection. To show how cumulative selection could speed up the process, he then took one simple phrase from Hamlet:[51]

---

[50] Dawkins, *The Blind Watchmaker*, 45ff.

[51] William Shakespeare, *Hamlet*, Act 3, Scene 2.

Methinks it is like a weasel

So, he wrote a computer program to randomly generate 28 capital letters and spaces. The first random sequence generated was:

WDLDMNLT DTJBKWIRZREZLMQCO P

Repeating with new random sequences, the results were no better, so he gave up on a straight random sequence approach. He then took the first sequence and made multiple secondary sequences from that set USING A FEW mutations—that is, he replaced a few letters/spaces. He then compared the new sequences with the target phrase METHINKS IT IS LIKE A WEASEL (all in caps, to make it easier) and selected the best new sequence. The winning new phrase was:

WDLTMNLT DTJBSWIRZREZLMQCO P

Then he generated a new mutation of that phrase, always selecting the best results. After 20 mutations and selections of the phrase, the computer achieved the phrase:

MELDINLS IT ISWPRKE Z WECSEL

After 30 manipulations, it was:

METHINGS IT ISWLIKE B WECSEL

In only 43 passes, he created the target phrase:

METHINKS IT IS LIKE A WEASEL.

That wasn't hard, was it?

But is there a fallacy? Admittedly, according to Dawkins, evolution doesn't know the desired final answer. Evolution just keeps improving and improving by selection. Richard Dawkins knew the target phrase. He intervened and used his intelligence (inherent in his computer program) to select the second-generation phrase to best match the final desired product. Each new selection of the previous mutated sequence is always compared to the known final target phrase.

In real life, the cell cannot know what the target alignment of codons along the RNA strands should be. There would be nothing to work toward! The direct oversight according to Dawkins had to be improved survival. But what is lacking in this analysis is the amount of time needed for that process. If one has the information as to the desired endpoint as Dawkins had, then the time is really inconsequential, and the desired result is quickly achieved. But this is far from the case. Dawkins' concept that if given enough time anything can happen simply fails because there is not enough time. Douglas Axe, a researcher at the Centre for Protein Engineering in Cambridge, England, has shown that the likelihood of transforming an enzyme to be improved over an existing enzyme is 1 in 100 thousand trillion trillion

trillion trillion trillion trillion (10^{77}).[52] Survival as the mode of selection is a slow, slow process—a particular creature would at least have to be born and reach the age at which they reproduce (having during their life developed a positive mutation).

As an example, while Dawkins may be right that certain cells may have become sensitive to light, helping them survive and develop many, many improvements until they could have a first-class eye. This certainly would have taken a long time, with each generation having to be born, develop a mutation which improved sight, then produce offspring with the new trait. Then, just imagine, not just one species randomly developing eyes and having greater survival, but many, many other species doing the same. At the same time, many other unique characteristics were also simultaneously moving down the evolutionary production line. To use a term Dawkins utilizes, I have a hunch that survival isn't the source of oversight which sped up the development of these new characteristics.

Dawkins' attempt to show how the evolutionary process could be sped up with oversight—that is, with cumulative selection based on survival, does depict a possible methodology for shortening the evolutionary development cycle.

---

[52] D.D. Axe, "Estimating the prevalence of protein sequences adopting functional enzyme folds," *J Mol Biol* 341(5):1295-1315 (2004), 8.

However, it is just not fast enough, considering the life cycle of a lifeform to always meet the historical record of the development of species. More research is need in that area.

Theists would ask, "Could the overseer be God directly involved in the process, or could the overseer be God in creating the physical constants so that the life would eventually necessarily be created in just the right fashion?" I'll just ask and leave it at that for now. But who else could serve as overseer? So, let's remember the Argument from Personal Incredulity when you think about responding to these questions. Be open to where the evidence leads you!

DNA Evidence for Evolution

After all that, I hope you don't think I am down on the evolutionary process. Why? Because even though it certainly appears to be true that *Homo sapiens* did not evolve from a single anaerobic cell in a slow gradual process, this does not mean there were not periods where slow gradual, cumulative processes resulted in an evolutionary path of development. Evolutionary biologists have been researching and informing us for over 150 years that lower forms developed into higher forms. They do have evidence for such. How the development of eyes or even different species came about still has many gaps which the future may or may not fill in as research continues. But there is strong evidence that an evolutionary process certainly was a part of the development of each life in the world as we see it today.

This concept seems alien to us (not meaning the out-of-space alien). I remember as a child hearing adults say, “I certainly didn’t come from an ape!” Most of these same adults didn’t say, “All creation occurred in six days!” as others may have believed at the time. They pretty well understood that the writer(s) of Genesis was/were trying to explain the essentials of Genesis, not the science of Genesis. And, as far man and woman’s creation, it was somehow different. They intrinsically knew that we are different from other animals. Those differences we will discuss in chapter 8. However, the BODIES of men and women are not very different from other animals. And how do we know this?

The answer is DNA. All life comes from pre-existing life (except apparently the first life). Basically, nearly all life has DNA. DNA, as discussed in chapter 5, contains all the information a cell needs to make its necessary proteins. How many letters are in a DNA strand? A lot! DNA can have 25000 nucleotide bases in its strand.

The Human genome (or the total of all the DNA information) is approximately **3.2 billion** base pairs in size in each and every cell. The DNA strand is about 6.6 feet long and, as indicated in chapter 6, is 40000 times thinner than that of a human hair. When the cell is about to divide, proteins attach themselves to the DNA. DNA plus its attached proteins are what we call chromosomes and can be seen under microscopes. Each chromosome contains but one DNA strand. The human genome, consisting of all of a person's DNA information, contains, in effect, the book describing you. The human genome is composed of the information from 22 chromosomes, plus the X chromosome and the Y chromosome. Compare that to the largest novel ever written (Marcel Proust's *In Search for Lost Time)* with a mere **9.6 million** characters. A thousand times different! That is quite a difference. But what is amazing, DNA from all living creatures is very similar. Here is a comparison.[53]

- Humans share 99.9% of the DNA sequences with each other
- Chimpanzees share 98.8% of the DNA sequences with humans
- Mice share 90%
- Dogs share 84%
- Chickens share 65% (even though non-mammal)
- Bananas share 41% (plant)[54]

---

[53] Lori Garrett-Hatfield, "Animals That Share Human DNA Sequences," at Education Seattlepi, available online at education.seattlepi.com/animals-share-human-dna-sequences-6693.html.

[54] The Animated Genome, available online at unlockinglifescode.org/media/animations/659#660.

Let's further examine the comparison of *Homo sapiens* and chimpanzee genomes. (See Figure 7.3). The Chimpanzee's chromosomes and Man's chromosomes appear nearly identical, except that the chimpanzee has 24 pairs, and we have 23. Our second chromosome is much longer than the corresponding chimpanzee's chromosome. However, if you attach the chimpanzee's chromosome labeled 2A in Figure 7.3 on top of its chromosome labelled 2B, it matches our #2 Chromosome.

Pretty remarkable.

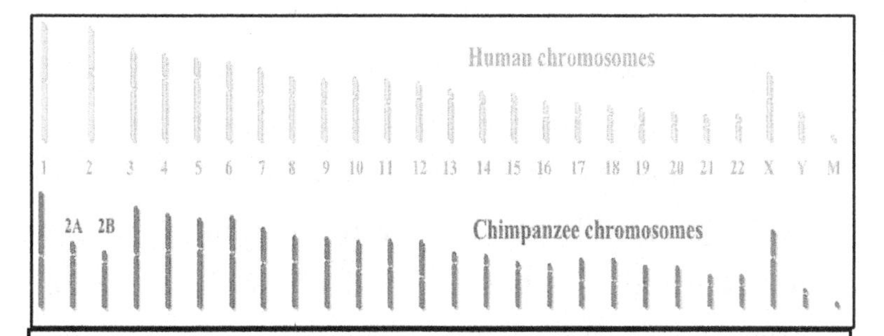

Figure 7.3 The human and chimpanzee chromosomes. Note that when the chimpanzee chromosomes labeled here as 2A and 2B are stacked on top of each other, they match the *Homo sapiens* chromosome pattern.

Indeed, an analysis of the base pairs of our second chromosome base pairs and the combined chimpanzee chromosome shows the chromosomes are essentially the same. Could the chimpanzee's extra chromosome 2B somehow have been attached to its second chromosome 2A during meiosis when the cells prepare themselves for reproduction, resulting in a new primate that we call *Homo sapiens*? Yes, there is evidence for that. Each chromosome has an end cap at both ends. So would the chimpanzee's 2A and 2B chromosomes. When one looks at man's Chromosome 2, it has four endcaps—two at the ends and two abutting near the middle. The top part of our Chromosome 2 is nearly identical to the chimpanzee Chromosome 2A, likewise for the bottom part of man's Chromosome 2 and the Chimpanzees 2B. Why this happened, we do not know. And, by the way, it doesn't mean that a *Homo sapiens* popped out of a female chimpanzee's womb. That 1% difference does make a difference. Imagine that! They both probably shared a common ancestor. When the split occurred neither the chimp looked actually like the present-day chimp nor did the man look exactly like the present-day man. However, the question of importance is, "Was it simply an accident, or was it an accident destined to happen?" Science cannot tell us, because that is a WHY question. For now, let's just say, "We do not know."

We do know, all of life is tied into the initial sequence which began with the first cell and the Origin of Life. We share our life with all living things through the sharing of the bases found in our DNA. Does that mean we come from a single tree of life, and all life forms are just different branches? Well, probably! To think otherwise would mean that different cell lines began life using the same identical sequences. But at this point, why force it?

Let's review a few things we have come to appreciate:

- Foremost, we need to look at all sides of an issue before we make judgments as to what is true and not true.
- The Universe is incredibly extremely specific in its laws and its constants. It is well-ordered.
- No one can specifically state how the Universe began from nothing (or how the Multiverse arose from nothing).
- As far as we know, the Earth is the only planet uniquely prepared to support life. Odds are that we shouldn't exist.
- Life arose in a completely unknown way. Many have theorized but no theory has produced Life. Life began from no Life. Odds are, we shouldn't exist.
- Once life began, it appears that, due to our genetic sequences being so similar, that Evolution does play a role in life's makeup.

Is the hand of God overseeing the process? Is the development of life and life's ecology, so utterly complex, just an undirected result of all those forces established at the beginning of time? One after another unlikely events continue to happen. Science is amazed at the complexity of it all.

But, in the end, rather than discussing what and why something happened, we are left with the term LUCK to answer the great questions of the origin of the universe and of life; yet, Divine Providence is left out. The question still remains, "Is man really just an ape?" The partial answer from a partial view of Reality seems to be YES, but only in the most limited way. The full answer is NO.

At the end of chapter 4, I mentioned the superb book *A Fortunate Universe,* written by Lewis and Barnes, which deals with the physics side of reality. However, when this wonderful book approaches the God question, it deals only with physicalism versus theism. It only really addresses the issue that fine-tuning truly exists. One could, in essence, just conclude we are so lucky OR God's providence provided for us. Many will just conclude that we are so lucky! The authors totally ignore the question of the FULLNESS OF REALITY. They do not pursue the evidence of Reality beyond physicality, for if a non-physical reality exists, then science will be seen as a wonderful but limited pursuit.

As of yet, we, too, have looked at only the physical side of Reality. We have looked solely to the point of a man's physical development. We have pursued only how the body evolved. In the next few chapters, we shall see evidence of the spiritual side—that man is not merely a body. Man is Body and Soul. The soul does exist and is not physical (just as information is real and is not physical). Those who study only the physical side will never see the FULLNESS OF

REALITY. The soul is transcendent to the physical, and it is the transcendent soul that makes us different from the other animals. The soul allows us to seek all the good things of life—the things that science cannot touch. The soul is what gives us free will, the option to do things that are beneficial or not beneficial for us, the option to create music, art, literature, the ability to understand the abstract, and the option to love or to hate. Even the option to dance, for, as we all know, that without soul, you can't dance! Shouldn't that be proof enough? So, let's look at the rest of reality. On to the next chapter!

Things to Think About

1. We hear about physical disasters such as earthquakes, volcanic activity, hurricanes, ice ages, global warming, etc. Has looking at the past disasters changed your thoughts about the development of life?

2. What are some evidences you have noticed that support Biologic Evolution?

3. Because some biologic developmental steps seem to follow evolution, does that necessarily mean all development is evolution

Chapter 8
Just Who Do We Think We Are?

There is a power struggle going on that has been around for thousands and thousands of years. It is a struggle to determine just who we are. Are we creatures, and therefore subservient to a creator to whom we owe our very existence, or are we each individuals, existing solely due to chance, able to do what one wills? In either case, while we are individuals, we cannot pretend that others are not linked. Mankind is a social entity. By man's nature, we are inclined to do things which benefit ourselves, but we also do things which benefit others, with no apparent gain for us. Why is that? We often freely and intentionally give up, to some extent, our freedoms (such as stopping for a red light) for the overall benefit of society because this act can and will benefit oneself in the long run. Is this a selfish act because life is much easier when we all just get along, or do we do good to others altruistically because we recognize others' true value as one of God's creatures? How should we live? Is morality simply situational, or are there God given precepts to follow? Your approach to life will depend very much on your understanding of the God question—"Are you a creature or not?"

So, let's begin with the "Who am I?" question. Scientism will simply state that we are just a collection of molecules with no God necessary. The Christian religion will state that we are created in God's image and are of infinite worth. Obviously, they both can't be right. This question can be restated in a somewhat conceptually different way which will lead to the answer of whether God exists. The substitute question is not whether God exists, but rather, is man **a body**—a single, unique unit composed of various molecules combined in a body which mechanically allows one to think and act (a monism), or is man **a body** conjoined with a non-physical **soul** (a dualism) allowing the transcendent to interact with the world? If man is body and soul, then the whole Neo-Darwinian concept of man as simply an entity originating through statistical interactions and mutations of molecules, with cumulative selection, resulting in survival for those who are the fittest or the cleverest, simply falls apart or should be recognized as a partial truth. Once the door opens to the existence of a spirit, then one also opens the door to survival of one's spirit after bodily death and to the possibility of one's spirit communing with a supreme spirit after death in a "heaven" or separated from that spirit in a "hell."

So, what are the ramifications if man is a monism? If man is just a body, he is simply a collection of molecules as the materialists and reductionists say. Mankind just happens to have a larger brain than most, a better arranged brain than other animals, a brain that can analyze the world around us which, thereby, allows us to draw conclusions and make decisions. We would be otherwise no different from animals. This raises a number of questions. Should we have dominion over animals? Can we kill other animals (even mosquitos)? And what about plants? Plants have molecules. Plants have life. Do we have the right to take another's life, even if that life belongs to a lower animal or to a plant? Of course, the materialist will say there are no rights. After all, in Darwinism, it is the fittest that survive. There is no morality other than the morality that if we don't rob, kill, or otherwise hurt another, then we will be able to get along and survive as a society better. Survival and reproduction of the species is what counts.

This sounds kind of harsh. Do materialists really not see any transcendence of man over other animals? In discussing suffering, prominent atheist scientist Richard Dawkins states in his book *The God Delusion* that the pain involved in killing a baby in utero during an abortion is really no different from the suffering involved in slaughtering a cow.[55] Actually, he doesn't call the fetus a baby. He calls it a late embryo with a nervous system, which apparently sounds more

---

[55] Richard Dawkins, *The God Delusion* (London, UK: Black Swan, 2016), 336.

palatable to him. I call it a baby because I know that an 8th month fetus in utero is no different than a premature baby *who* (not *which*) is born in the 8th month of pregnancy. The act of leaving the womb by abortion or by, say, a C-section, does not make one a lump of cells, versus a human baby filled with life. Killing a baby and killing a cow are not equal events because the "essences" involved are not equal. It is also clear in Dawkins' world that there is no special meaning of life. Certainly, for Dawkins, there is no life after death, for when the body is dead there remains nothing other a collection of chemicals.

The dualist view is much different. Because man has a soul which is transcendent, man has a greater worth than a cow which has only instinctive concerns. A cow never asks itself, "Was that the right thing to do?" as a human might. Necessarily, since the soul is not physical, it can continue its existence after the body has died.

So, is there any evidence that we are body and soul or just a body? Our life experience suggests that we are body and soul. In our daily lives, we often speak as if the body and soul are two different entities. For example, we speak colloquially of the mind's being different from the brain. When we forget something or perhaps when we do something stupid, we say, "I've lost my mind!" (or else someone else will say it for you!). We never say, "I lost my brain." It certainly would truly be a sad state of affairs for us if we didn't know where our brain is. But the feeling that your body and soul are distinct, yet somehow conjoined, is not just a colloquial form of speech. I can remember my mother saying to me as she entered the twilight of her life that she felt like she, the person, was trapped inside her body. She wanted to do so much more, but her physicality was limiting her. She felt free and wanted to run free, but that was no longer a possibility for her. In my mind (not brain) when she told me this, I simply said, "Yeah Mom." I was not very understanding. How sad. However, as I have aged, I have begun to understand what she meant. My body doesn't work with my person as it used to. I desire to run fast, jump high, memorize a list of ten things with little effort, etc. My spirit has not dulled, but my body has. My body says, "No way!" In addition, my body often says, "I need energy!" and I say to myself, "Energy! I just fed you a Big Mac, an order of fries, and a Coke." That used to work, but not anymore. I talk to myself as if the body is some third party. My personal experience is that there is a separation, a distinctness between body and soul. With aging, my body is gradually beginning to become disjointed from my soul. My youthful spirit is trapped within an aging body. I am not alone in this experience.

The concept of the soul is an ancient one. One of the earliest references goes back to Homer (late 7th, early 8th Century BCE) and the epic poems—the *Iliad* and the *Odyssey*. The Homeric concept was that the soul is something that a human being risks in battle and loses in death. It is what departs from the person's body at the time of death, and travels to the underworld, where it is simply a shadow of the deceased person.[56] Homer's concept was solely related to human life and is not connected with life in general. Homer's soul was immortal. By the 6th century BCE, the term soul was applied to all living things and began to be applied to non-physical activities, such as to satisfy one's soul by eating rich foods, sexual desire, etc., and to degrade the soul by dishonor.

Plato further detailed the use of the word soul when in the *Republic* he has Socrates ask, "Haven't you realized that our soul is immortal and never destroyed?"[57] In the *Phaedo,* Socrates discusses whether after death the soul "still possesses some power and intelligence."[58] According to T. M. Robinson in his book *Plato's Psychology*, Plato's human souls are intelligible, partless, and imperishable.[59] Other Greek theories of the soul, such as that of Epicurus, included thoughts that the soul is a body within a body, like atoms of

---

[56] Hendrik Lorenz, "Ancient Theories of Soul," *The Stanford Encyclopedia of Philosophy* (Summer 2009 Edition), Edward N. Zalta (ed.), (608d), at plato.stanford.edu/archives/sum2009/entries/ancient-soul/.

[57] Lorenz, "Ancient Theories of Soul."

[58] Plato. *Plato in Twelve Volumes*, Vol. 1 translated by Harold North Fowler; Introduction by W.R.M. Lamb. (Cambridge, MA: Harvard University Press, 1966).

[59] T. M. Robinson, *Plato's Psychology* (Toronto: University of Toronto Press, 1995), 29.

some unique substance. Rationality is contained as one part of this substance while nonrational parts receive sense impressions. Pleasures and pain of the soul belong to the rational, and the animation of the body to the irrational.

Confused? Don't worry. The gist is that from earliest times the greatest thinkers of antiquity have recognized a separation from bodily functions and rational functions. "But aren't we all just a collection of atoms and molecules?", some modern-day atheists might ask. "Thoughts are just electrical discharges across synapses of the brain of human bodies—bodies which have the ability to replicate themselves through sophisticated forms of cell division. To be sure, we atheists understand that the ancient philosophers were extremely intelligent, but to be honest, they just didn't have the knowledge that modern man has." Who's right and who's wrong? Both can't be right, or even partially right. So, there is a standoff—those who believe that mankind is body and soul versus those who believe only in the body. It would be nice if there were evidence one way or another. But, ah, there is documented evidence of the separation of the body and a soul.

Near Death Experiences

One answer lies in examining Near Death Experiences (NDEs), which is something many atheists seem to avoid. They might say, "It's not possible, so why waste your time?

The brain is undergoing great stress and is not thinking rationally." (Remember the Argument from Incredulity—because one doesn't personally feel something is possible, by itself, doesn't mean it's not.) Ignoring data is not a scientific approach at all. It would be an example of willful blindness. There are now thousands of NDE cases to be examined. The term Near Death Experience is really a misnomer. By all accounts, the body actually does die. If the body didn't die, then the event would not be included in the studies. Perhaps one should call it a Near Permanent Death Experience. But it is what it is, and I am not going to rant and rave for a new name. As it is, many scientists have investigated and reported these cases in numerous articles in celebrated scientific medical journals, such as *JAMA* (*Journal of the American Medical Association*) and *Lancet* (a prestigious British medical journal). So, let's see what these studies have found. Do they show a soul component independent of the body?

Let's start with, "Just what are NDEs?" Near death experiences are human experiences in which an individual actually clinically dies. The individuals, though clinically dead according to a lack of brain activity or other defining characteristic, suddenly return to life. They recount a fairly common story—the person leaves the body, floats around, and sees one's own body. Here the story has variations. Sometimes the person may float into another room, perhaps a waiting room outside an operating suite. There, they may hear conversations, which they later report. They will see and be able to identify clothing being worn by the people in the room, even the clothing of people whom they do not know. They have emotions. They may not like what is being said. Thus, they have memory, intellect, and emotions, and they can receive communications—it seems as if they truly are existing outside the body.

Other times, the person may see a light and head into the light ("Go to the light," the spiritual medium, Tangina Barrons, said in the fictional 1982 *Poltergeist* movie.[60]) There they may meet a deceased loved one, or Jesus, or people they have never met before but later identify in family photos. Others travel to the far reaches of the universe, proceed backward in time, and see colors they have never seen before or musical tones they have never heard before, etc. Eventually, they return to their body, and the body begins functioning again. Often, one is given the choice to return or not, or is told it is not one's time. In other instances, one simply just returns to the body.

All this sounds utterly fantastic. Some would say psychotic. It does not match what we seem to know about ordinary life. But before you judge, just keep remembering the examples of relativity, or quantum physics, or gravity waves, etc., all which do not make sense according to our everyday life. And, since NDE reports "seeing" and "hearing" without

---

[60] *Poltergeist,* produced by Steven Spielberg (MGM Entertainment, 1982).

using the eyes and ears, why wouldn't colors and sounds be so much more spectacular, for our biological eyes and ears are limited to a narrow range of frequencies of light and sound which can be seen/heard. Indeed, research shows that we do not see all that is to be seen or heard. Without the restriction of physical eyes and ears, we should be able to see much more of reality. After all, there is more in our physical world than we are aware of. For instance, the light seen by bees is much different from what we see, and dogs can hear at frequencies much higher than what we can hear.

History has much to say about spiritual reality. Out of body experiences have been reported throughout history, especially modern history. To believe there is nothing to this, one has to believe all those people in the past, including intelligent, highly educated individuals, were liars, mentally ill, or simply confused. I personally believe that a preponderance of future studies will eventually convince nearly everyone that we are body and soul. Like the physicists in chapter 2 who had to become open-minded and change their view of classical physics because they could no longer deny the reality of phenomena which only quantum physics/relativity, atomic physics, nuclear physics, elementary particles, astrophysics, etc., could explain, modern scientists will need to change their view of a soulless body and accept the body and the soul model.

So, where's the evidence?

There are many reports of NDEs. There is even now a *Journal of Near-Death Studies*, a peer-reviewed scholarly journal (ISSN 0891-4494) devoted exclusively to the field of near-death studies. When I last checked, there were over 900 books on the subject. Since not everyone is looking at this topic with a cold scientific eye, I will refrain from discussing cases, though perhaps true, which have no verifiable evidence to examine. The cases I will discuss truly have evidence associated with earthly events which the individual could not know of, could not have seen, heard about, etc. These NDE cases involve situations in which the person could not have possibly known ahead of time the facts they relate when they return to life and report the NDE. All of these cases would absolutely require the use of personal attributes such as seeing, hearing, feeling, emotions, intellect, memory, etc.—functions which normally are controlled by the brain. Only the brain was not there, instead remaining in the person's body on an operating room table, or, perhaps, at some site of a serious accident, or wherever.

What is particularly interesting about reported NDE events is the extreme lucidity of the encounter. This lucidity contrasts with experiences everyone has in everyday dreams, which always have an unreal aspect to them. Many a physician has tried to explain NDE by saying the NDEs are produced by brain chemistry. However, Dr. Peter Fenwick, a neuropsychiatrist and a leading authority in Britain concerning NDEs, in discussing the brain of a patient having a NDE states:

> "The brain isn't functioning. It's not there. It's destroyed. It's abnormal. But yet it can produce these very clear experiences ... an unconscious state is when the brain ceases to function. For example, if you faint, you fall to the floor, you don't know what's happening and the brain isn't working. The memory systems are particularly sensitive to unconsciousness. So, you won't remember anything. But, yet, after one of these experiences [an NDE], you come out with clear, lucid memories ... This is a real puzzle for science. I have not yet seen any good scientific explanations which can explain that fact."[61]

The reporting of events outside the trauma site, which would be completely unknowable by an individual who is dying, is another case which is impossible just considering brain activities. How can the brain "see" outside the room it is in? Many of the arguments against NDEs are simply hand-

---

[61] Kevin Williams, "People Have Near-Death Experiences While Brain Dead," available online at www.near-death.com/science/evidence/people-have-ndes-while-brain-dead.html

waving—speculative arguments that do not examine the cases in detail.

Unfortunately, on the other hand, due to the nature of people in the world, in reading NDE accounts, one has to remain skeptical to a certain extent as there might be a shyster about. Yet, even if only one of these cases is true, we have evidence that having a body is not necessarily a requirement for being a person.

So now, let's learn more about the NDE experience.

The Typical Characteristics of NDE

If NDE's happen as frequently as I suggest, one would think there might be consistent patterns in the NDE experience. In fact, there are. Over and over again, people who have experienced NDEs have reported common trends. The statistics on these trends have been collected and reported by many individuals and research groups. The following information has been reported by Dr. Jeffrey Long.[62] He has identified twelve common elements found in Near Death Experiences and the percent of NDE'rs who have experienced specific elements:

---

[62] Jeffrey Long, and Paul Perry. *God and the Afterlife: The Groundbreaking new evidence for God and near-Death Experience* (HarperOne, 2016), 11(ff).

1. An Out-of-Body Experience (74.9%)
   A person's consciousness exits the body, sees the body, below and can move on.
2. Heightened Senses (74.4%)
   A greater awareness, viewing colors never seen, hearing sounds never experienced.
3. Intense and generally positive emotions or feelings (77.7%)
   Incredible feelings of peace and love.
4. Passing into and through a tunnel (33.2%)
   A sense of moving through a tunnel.
5. Encountering a mystical or brilliant light (64.8%)
   Intensely bright light with no discomfort, the feeling of uniting with the light.
6. Encountering deceased relatives/friends or mystical beings (57.8%)
   Sometimes even relatives whom one has never met before.
7. A sense of alteration in time and space (60.5%)
   Time is non-existent, instant transfer from point to point in space.
8. A life review (21.8%)
   All the moments of life seen in an instantaneous review.
9. Encountering Otherworldly (heavenly) Realms (51.0%)
   A reality like none ever experienced.

10. Encountering or learning special knowledge (57.6%)
    Can be earthly or spiritual, includes knowing everything about the universe.
11. Encountering a boundary or barrier (31.0%)
    A boundary which NDE'rs understand is a point of no return.
12. A voluntary or involuntary return to the body (58.5%)
    Sometimes NDE'rs are told this is not their time, others are given an option.

Sounds like a wonderful "place" to be in, assuming it is not just a comatose response as some scientists suggest. Also, note how many occurrences are consistent with historical descriptions of heavenly experiences. This is not a modern experience. It has been happening for a long time.

Now let's look at some cases and "hear" the accounts.

Case 1: The "dentured" patient

This is the well-known case of a Dutch man who in 1979 had a myocardial infarction on a cold day while walking along a road near the village of Ooij in the region of the city of Nijmegen, The Netherlands.[63] The man was taken to a nearby hospital. He was clinically dead as seen by the patient's corpse-like blue and blotchy skin, as well as the wide-open nonreactive pupils, blue nails, and lips typical of a person with severe oxygen starvation. He was ice-cold to the touch, was unconscious, and did not react to anything; and he was still in ventricular fibrillation. When trying to intubate, they discovered the man had dentures and these were taken out by the male nurse and placed on a shelf on a nearby cart before they started. The ER team worked diligently although they felt it was hopeless and after an hour and a half, they were able to restore the heartbeat.

Afterwards, the man fully recovered but still was missing his dentures. He saw the male nurse, recognized him, then asked where the dentures were. The nurse was shocked, and the patient told him that he had recognized him because he was floating above his body and saw him working on him, and he saw the male nurse take out his dentures and place him on the wooden shelf of the crash cart.[64] (Note: *The Lancet* article reported the dentures being placed in metal

---

[63] P. van Lommel, R. van Wees, V. Meyers, and I. Elfferich, "Near-death experiences in survivors of cardiac arrest: A prospective study in the Netherlands,"

Lancet 358, (2001), 2039–2045.

[64] Rudolph Smit, *Journal of Near-Death Studies* ndst-27-01-03.3d 24/10/08 12:58:35 47.

drawer on the crash cart.) The man also described the other items on the crash cart, the personnel in the room, and various other equipment.

This story was published in *The Lancet*, a prestigious British medical journal. While the detail recalled was quite remarkable, some claim that even though the man appeared comatose he was still cognizant due to his hypothermic condition.[65] However, the possibility of the patient being able to see and remember all the surroundings while in a comatose condition makes it unlikely for that to have occurred.

Case 2: The Shoe on the shelf and the visit to the waiting room

Kimberly Clark Sharp was a social worker working in Intensive Care and Coronary Care at the Trauma Center at Harborview Medical Center in Seattle Washington in 1977. One April day in 1977, a Spanish-speaking woman named Maria was admitted unconscious into the medical center and placed in the ICU. Kimberly met with her the next day. Three days later, the woman suffered a cardiac arrest. Kimberly happened to be on the floor at that time and witnessed the relatively easy resuscitation.

---

[65] Gerald Woerlee, *Journal of Near-Death Studies*, 28(4), Summer 2010 IANDS.

Later in the day, Kimberly was called to speak with the patient who was very agitated. The patient described herself as having floated above her body and seeing her body on the bed. Her spirit then went outside the window and over to the Emergency Room entrance. She described correctly the driveway and other details of the Emergency Room entrance, including people entering the area. Then she mentioned that she saw a blue sneaker on the ledge of the building three or floor stories above the ground. She even mentioned some details as to how the shoe was positioned, etc.

At that point of her life, Kimberly was unfamiliar with the studies of NDE. This oddity perplexed Kimberly. She looked out the window and saw that Maria could not have seen the Emergency Room entrance because of the presence of a roof blocking the view. In addition, there was the odd comment about the shoe. Kimberly just had to go check the shoe issue. Maria was on the second-floor north end of the Medical Center. No shoe was there. Kimberly became determined to find out if the shoe existed. She did eventually find the shoe on the 3rd floor ledge of the west end of the Medical Center by leaning outside the window above the ledge. She later confirmed that the sneaker could not have been seen from the ground below unless one was at least a half block away.[66]

This account is just too bazaar to have been made up. Physical evidence of the shoe exists today. Furthermore, this event changed Kimberly's life. While maintaining her social work occupation, she began to give talks on the subject, and even founded, in 1982, the Seattle International Association of Near-Death Studies, the world's oldest and largest support group for near-death experiencers.

Can the Blind See?

There are many cases of the blind who, while experiencing an NDE, are suddenly able to see. If one believes that the soul can communicate to other persons, and one believes that the soul unites with the body at conception then uses the body's faculties during everyday life, then it would be reasonable that the soul once outside the body would be free of the body's limitations and still be able to see, hear, think, have memory, and even emotions. This would also be true of blind people. Once free of the defective visionary system with separation from the body at death, the blind would

---

[66] *Kimberly Clark Sharp,* The Other Shoe Drops: Commentary on "Does Paranormal Perception Occur in NDEs?" *Journal of Near-Death Studies* Volume 25 (2007), *245-250.*

suddenly be able to have vision. Let's review a few cases where this seems to be the case.

Case 3: The Blind See

Dr. Kenneth Ring in his book *Mindsight: Near Death and Out-of-Body Experiences of the Blind* recounts his efforts to determine if the blind actually do see their surroundings once the consciousness leaves the body. He presents the results of his studies and discusses many cases of the blind who see clearly while experiencing their out-of-body/near death event. In one of the cases, he relates the account of a woman who went into the hospital in 1991 for a one-hour biopsy procedure. Unfortunately, due to a medical mishap, the Superior Vena Cava was cut during a chest biopsy, then inadvertently sewn shut—shutting off normal blood flow from the brain and resulting in a swollen head due to the blood backup. Six hours later while still in recovery, she awoke and said "I am blind! I am blind!" She soon stopped breathing and was manually being forced to breathe by personnel using an ambo bag—a bag shaped like an American football which is squeezed in order to force air into her lungs. Not knowing that the Superior Vena Cava was accidently sewn shut, they rushed her to an imaging suite to do an MRI to see what was going on. At an elevator along the way, she suddenly left her body, "stood" outside the cart and began to watch the happenings. At this elevator, while her body lay flat on the gurney, numerous tubes, the ambo bag, plus hospital personnel (three individuals on the left and three on the right) blocked any possible lateral line of sight, her consciousness (soul) saw clearly her boyfriend and another man—the father of her eight-and-a-half-year-old son—standing down the hallway 15 to 20 feet away. She saw them in shock as to what was going on, for, after all, this was to be a simple biopsy. It was then that she became enveloped by a bright light. She was urged to return and eventually agreed to it. At 7:30 pm, about ten hours later, she was operated on to open the Superior Vena Cava, but it was too late, and she became permanently blind. Her story about the gurney, the elevator, and the two men was corroborated by hospital personnel and by the two men who stated that indeed they were there. They all agreed she was not in a position to have seen the men or to have described the gurney and personnel in such detail. Certainly, the clarity she described just was not possible.

The lucidity of her account defies the arguments of medical doctors who claim the memory is just the result of oxygen deprivation, brain stimulation due to trauma, drugs, or whatever. As anyone who has had a dream or has undergone anesthesia, the brain is incapable of creating a truly reasonable story. Think about it. What dream have you ever had that showed truly rational, everyday behavior? Here, the woman had become blind. Yet she could describe the whole sequence of events at the elevator in a clear concise manner.

Her body's eyes could not see, and her hearing could not have given the visual details. There is no means by which a blind person could describe such events.

Is it just a Dream?

It turns out that there are many reported cases of blind people having NDEs and OBEs (Out of the Body Experience, but not necessarily a Near Death Experience) being able to describe their particular events with visible detail. From studies performed by Dr. Kenneth Ring, 77% of those blind individuals who had an NDE or an OBE could with unhesitating declaration, in fact see.[67] (In this context, "seeing" does not mean seeing with their damaged eyes; rather, it means receiving visual images without using their biological eyes.)

To understand what vision means to a blind person, there have been studies evaluating a similar but different imaging which might occur for the blind—that would be dreaming. Let's review the research results on the dream characteristics of the blind and see if the vision of the blind due to NDEs and OBEs just be vivid dreams? So, what is the normal imagery in dreams of the blind? According to Dr.

---

[67] Kenneth Ring, and Sharon Cooper, *Mindsight: near-Death and out-of-Body experiences in the blind* (Kearney, Ne: Morris Publishing, 1999), 122.

Donald Kirtley as summarized in a 1975 report on the dream imagery of the blind:[68]

1. There are no visual images in the dreams of the congenitally blind.
2. Individuals blinded before the age of five have no visual images in their dreams.
3. Individuals blinded between the ages of five and seven may or may not have visual images in their dreams.
4. Individuals blinded after the age of seven may see visual images in their dreams, but the images may fade with time.

This information is important to note, for those who became blind early in life, they do not have dreams with images, so, if such an adult has a NDE or OBE later in life, the images must be new data to them and their consciousness. It is clear that the images seen cannot be dreamlike in nature. Let's look at another case involving a blind person.

---

[68] Donald D Kirtley,. *The Psychology of Blindness* (Chicago, IL: Nelson-Hall, 1975).

Case 4: Carla's Story

Carla was born prematurely and like many premature babies in the 1950's, she was placed in an oxygen filled incubator. Like approximately 50,000 other babies born at that time, the oxygen levels were set too high, which resulted in almost total blindness due to damage to the optic nerves. She was totally blind in one eye and could only see limited forms, but no detail in the other. When she was 42, she experienced an NDE while undergoing a hysterectomy. She then had an Out of the Body experience. Looking down, she was able to clearly see the anesthesiologist working with a respirator, herself on the operating room table, and the Telemetry screen on the ceiling. On the screen, she could see wavy lines showing her heart rate and other measurements but did not know at the time what those lines meant. She noted that the staff wore green, although at first, she didn't know what to make of the color. It was darker than white and less intense than red. Both the surgeon and the anesthesiologist were tall. The anesthesiologist had white hair, although he was really too young to normally have white hair. His hair was covered but not completely by a surgical cap. He had a moon-shaped face. She could see a quite concerned face and hear him say, "We are losing her!"

At this point, her soul was swept into a white room where she visually saw and auditorily heard all of her life experiences beginning with her childhood. Her vison through all this was crystal clear. During her life review she could she and describe various playgrounds and her high school building plus the associated colors. Eventually, she returned to her body. She later described all this to her doctors. [69]

Another pertinent question concerns whether the NDE and OBE experiences of the blind are similar to that of sighted individuals. In his research, Dr. Ring found the following NDE features.[70]

Frequency of Common NDE Features Mentioned in Interviews

Feature	# of Patients	Blind %	Non-Blind %
Felt peace, well being	20	95%	77%
Reported a sense of separation, from the body or an OBE	14	67%	75%
Saw one's own body	10	48%	----
Went through tunnel or dark space	8	38%	33%

[69] Ring, *Mindsight: Near-Death and Out-of-Body Experiences in the Blind*, 86.

[70] Ring, 38.

Met Others (spirit, angels, religious personages, etc.)	12	57%	58%
Saw a radiant light	8	38%	65%
Heard noise/music	7	33%	74%
Had a life review	4	19%	22%
Encountered a border or limit	6	29%	31%
Was given a choice or told to return	10	48%	59%

The percentages are not exactly the same, but they do show a range of possible interactions for the blind consistent with those observed by the non-blind.

Will I Meet You on the Other Side?

If one only believes in Scientific Materialism, then when you die, that's it. That's not comforting at all. I certainly hope that is not true. Certainly, the many creative authors, song-writers, poets, etc., have long supported the concept of life after death. Certainly, the history of all civilizations supports that concept. But if it is not true, then it is not true. Wishing it were true will not change the final outcome. So, is there any evidence that it is true? Has mankind just been making this up? Let's see if there is any evidence.

Case 5: Meeting the Extended Family

Let's just take the popular *Heaven is for Real*[71] story. As you may know, this story was made into a book, and then into a movie. For this reason, many would discount the story. However, those facts do not make it untrue. In addition, the story contains so many common elements of NDE accounts, that it is worth including. In this story, a 3-year-old boy named Colton has his appendix burst. This results in an NDE. In his experience, the young boy meets Jesus and is greeted by both his grandfather, who died before he was born, and a sister Sonja, who as it turns out, was miscarried before Colton was born and never was previously mentioned to the boy before the NDE. As such, he had no way of knowing these individuals. The grandfather is identified during a time when the parents are looking in a family album. Colton later questions his parents about the existence of an unknown sister, as he had met a girl during his NDE who told him she is his sister.

Now as stated, there are many such stories in NDE accounts of an NDE'r seeing persons whom one had never personally met previously during their life. Often the NDE'r is greeted by relatives, friends, and neighbors who have pre-

---

[71] Todd Burpo, and Lynn Vincent, *Heaven is for Real: a Little Boys Astounding Story of his Trip to Heaven and Back* (Nashville, TN: W Publishing Group, 2014).

viously passed on. In some cases, unknown to the NDE'r, the person who meets the NDE'r may have just died. Here is another case.

Case 6: The Unmet Sister

For instance, in one case, a girl suffering from encephalitis had a NDE and met a family friend who told her to go back. As it turned out, the family friend had died the day the girl entered the hospital. Later, in the same experience, the girl had another NDE event and met an individual who said she is her sister. After her recovery, the girl draws a picture of the "sister" she had met. Her parents asked her who the drawing was of and were told it was her sister she met when sick. The shocked parents' faces turned ashen. They left the room, only to return to tell her that the sister she had never known had been struck by a car and died before she was born.[72]

Case 7: O Death where is thy sting?

The Life Review is considered to be one of the most transformative and powerful aspects of the NDE. One man, named Roger, stated:

---

[72] Jeffrey Long, and Paul Perry. *Evidence of the Afterlife: The Science of Near-Death Experiences* (Harper One, 2011), 128.

"I went into a dark place with nothing around me, but I wasn't scared. It was really peaceful there. I then began to see my whole life unfolding before me like a film projected on the screen, from babyhood to adult life. It was so real! I was looking at myself, but better than a 3D movie as I was also capable of sensing the feelings of the persons I had interacted with throughout the years. I could feel the good and bad emotions I made them go through. I was also capable of seeing that the better I made them feel and the better the emotions they had because of me, the more credit I would accumulate and that the bad emotions would take some of it back just like in a bank account."[73]

Obviously, I could go on and on. For more cases, go to the websites for the International Association for *Near Death Studies (IANDS) or the Near Death Experience Research* Foundation (NDERF). The question which needs to be answered is, "How can this be?" Unfortunately, those individuals who promote the "we are all just a bunch of molecules" theory simply sidestep and address these issues superficially or not at all. They may accept the explanation that a person under morphine or some anesthetics have hallucinatory dreams, and, of course, many are not under the influence of drugs or anesthetics. We know that dreams,

---

[73] Long and Perry, *Evidence of the Afterlife: The Science of Near-Death Experiences*, 191.

much less hallucinatory dreams, are not lucid. And how can one report factually those details one could not possibly know?

By looking at the wealth of experience of NDE's, it is becoming more and more evident that the personhood of the individual resides in the soul, and that the soul united with the body utilizes the body functions to see, hear, smell, process, store information, etc. When the body parts fail or fail to develop, suddenly one cannot see, feel, hear, etc. With brain damage, the pathways used by the spirit fail to perform as they once did. Ah, but with death of the body, the soul leaves the body. The soul can communicate, feel emotions, think, and even has a free will which, if offered, can choose to return to the body or not. This is what the evidence tells us if one is open to accepting the other facet of reality—the spiritual side. And, if one's soul decides not to return the body, or the soul has no choice, where does it go? From the reports of the NDE survivors, they go to the light! For Christians, Jesus is the light of the world (John 8:12).

This distinction between physicality and non-physicality that we feel in our daily lives is what we have known all along when we say, "I have lost my mind!" The soul is the image of God. It is because of the soul that we can love, or hate, build or tear down, weep, or laugh. And because the soul uses the body, we must not neglect our body; it is our "temple" of our spirit. The wonderful, wonderful brain is a processor of information, and along with other body parts must be protected. Those with the highest IQ's indeed have the most efficient brains, but as we all know, it is not the brain (unless the mechanics of the brain interfere), but the soul that determines if one is a nice person to know. One must also develop and protect the soul, which obviously cannot be damaged physically but can be damaged by the selfish choices we make in life, that is by sin in one's life, or made holy by the graces one receives in life.

Still not convinced? Still believing that thousands and thousands of people with no thought of gain are reporting these NDE events in a logical consistent way, as being due to a hallucination, which is always not logical and consistent? It is not that those individuals who suddenly have a heart attack, were in a car accident, or were simply in the wrong place at the wrong time, had the time and thought to make this up and to observe in such details the world around them. Remember, this has been going on for thousands and thousands of years.

Evidence from the Medical Field

The materialists believe in monism, that we are just a body with a brain that functions as the central processor. The mind is just another moniker for the brain. However, is that true? We have just discussed NDE's whose stories are reasonably consistent with each other, and which tell the tale of the blind seeing as if with eyes. But is there any physical evidence that the mind and the brain are separate entities?

The 20th Century and following have been exciting times for the development of techniques to determine the functioning of the human brain. One of the early pioneers was a neurosurgeon by the name of Wilder Penfield (January 26, 1891 – April 5, 1976). Dr. Penfield received world-wide recognition for his work on epilepsy and for the mapping of the of various regions of the brain corresponding to the various body parts, such as hand, toes, eyes, etc. Dr. Penfield could do this work because he developed a technique, which allowed him during open brain surgery to electrically stimulate various locations within the brain while the patient remained conscious. Thus, the patient was able to communicate with Dr. Penfield while being stimulated with an electrical probe the desired region within the patient's brain. As squeamish as this makes me feel, it must be remembered that the brain which receives pain sensory inputs from all parts of the body, does not, itself, feel any pain!

In his studies, Dr. Penfield was able to not only correlate the corresponding location of body parts to the brain and cause a tingling in a patient's body part, say the right index finger, but was also able to elicit memories stored in the patient's brain. He was even able to generate feelings of hallucinations, illusions, and déjà vu (the feeling you've been there before—often felt by patients who suffer from epilepsy). Dr. Penfield received many awards for his work and was possibly the most noted Canadian citizen of his day.

However, unlike NDEs, Dr. Penfield was not able to create hallucinations of what people were saying at another point in space, like the next room. The memories and images he elicited were real things based upon the patient's various life experiences and knowledge. Thus, he could conclude that memories are stored physically within the brain, in a manner similar to, but of course different from, the way we store images on a hard disc in our computers.

Now, Dr. Penfield was also interested in the mind and its reasoning power. Is the mind and the brain the same? If this were true, then Dr. Penfield should have been able to locate a site or sites within the brain to generate reason. He should have been able to find a region within the brain that would be able to think abstract thoughts (to think outside the box so to speak). To think about a hula hoop and call it a circle is not the same as to think of a non-existent circle and what it means to have a constant radius, and to perhaps reason how to calculate the area encompassed by the circle. No matter how much Dr. Penfield stimulated various regions in the brain, no such abstract thinking occurred.

Dr. Penfield began his studies as a monist—there is nothing but a materialistic body with a brain. As he analyzed 1000's of patient studies, he began to change that thought. He came to the following observations:

- If the mind is found within the brain, then every once in a while, putting an electric current on, say, the cortex, should result in an abstract thought.
- Those who have seizures, whether general, or just focal, such as having a twitching of a finger, should at some time during one of these electrical stimulations have an intellectual thought, but they never do.
- As we all know, people are capable of lying. Lying is an intellectual activity and an act of the free will. If the mind is contained in the brain, then at some time, the electrical stimulus should generate a lying response from the patient. This never happens.

Dr. Penfield came to the conclusion that the mind is separate from the brain. There is a dualism of the person.

If you are cynical, you might just believe that Dr. Penfield and those who have followed him, have just missed that magic spot where the physical brain does all of its abstract thinking. Really? Think about it. But let's move on now because there is other evidence that we are more than molecules and that a being far greater than us exists.

Things to Think About

1. Could a robot ever be a person?

2. Could Near Death Experiences just be a chemical manifestation originating in the brain?

3. Our "vision" and "hearing" senses are limited by the earthly vessels we call bodies. What of our other senses? How might being outside the body enhance our other senses.

Chapter 9

What Kind of Science Is Con-Science?

The question as to whether we are body, or body and soul continues. Chapter 8 dealt with Near Death Experiences of thousands of people. Now we are going to discuss something affecting BILLIONS of people. That would be the Conscience. We all have one. The conscience doesn't control behavior, but rather the conscience is there to guide the individual. At times, we all sidestep our conscience, or gradually we may pervert our conscience and, as a result, what was once bad may be considered okay.

Figure 9.1 The Conscience By François Chifflart (1825-1901)

So, is our conscience a built-in molecular construct? Do our genes develop the molecules in our brains in order to alter the functioning of the brain, thereby affecting our decision making?

Those who believe in Scientism generally accept that a conscience exists, they may just not use the term—conscience. The reality of a conscience has a long history. The 5th Century BCE Greeks were aware of a conscience. They termed it "*suneidesis*"—the sharing of knowledge with oneself, usually a moral sharing of one's being in the wrong.[74] Typically, the conscience was seen not as a conviction, but rather as a "voice". The Romans had a similar concept. They used the word "*conscientia*"—"con" meaning shared with, "scientia" meaning knowledge.

The Greeks and Romans then considered the Conscience an inner communication, which sounds remarkably similar to the communications "heard" by those who experienced a NDE, without having any direct auditory input through the ears.

The Catholic Church in its decree *Gaudium et spes* (Joy and Hope) from the Second Vatican Council, promulgated in 1965, writes concerning the spiritual aspect of the conscience:

> *"In the depths of his conscience, man detects a law which he does not impose on himself, but which holds him*

---

[74] Stephen Darwall, "Moral Conscience through the Ages: Fifth Century BCE to the Present," at Notre Dame Philosophical Reviews (March 10, 2015), at ndpr.nd.edu/news/56313-moral-conscience-through-the-ages-fifth-century-bce-to-the-present.

to obedience. Always summoning him to love and avoid evil, the voice of conscience can, when necessary, speak to his heart more specifically: DO THIS, SHUN THAT. For man in his heart has a law inscribed by God. Obey it is the very dignity of man; according to it he will be judged."

"Hence the more right-conscience holds sway, the more persons and groups turn aside from blind choice and strive to be guided by the objective norms of morality. Conscience frequently errs from invincible ignorance without losing its dignity. The same cannot be said for a man who cares but little for truth and goodness, or for a conscience which by degrees grows practically sightless as a result of habitual sin."[75]

Here is the summary of the Catholic Church's teaching:

- Everybody has one.
- The Conscience has two functions:
  - Urges one to do GOOD and avoid EVIL.
  - Directs one as to WHAT IS GOOD.
- This Foundation has been laid by God but can be led astray.

---

[75] Pope Paul VI, Pastoral Constitution On The Church In The Modern World *Gaudium et spes* (1965).

- Our Consciences can become malformed or even further developed.
- One's actions will, however, be JUDGED depending on whether you followed your conscience or not, not whether the action is truly evil or good!
- We have an obligation to properly form our conscience.

The Conscience is non-physical in nature.

From Scripture, one sees the term Conscience is only used once in the Old Testament and 30 times in the New Testament. Paul in 1 Timothy 1:5 states "*the aim of our charge is love that issues from a pure heart and a good conscience and sincere faith.*"

The Conscience when "good" leads us to the greater good – LOVE.

1 Timothy 4:1-2 "*Now the Spirit expressly says that in later times some will depart from the faith by giving heed to deceitful spirits and doctrines of demons, through the pretensions of liars whose consciences are seared.*"

Hebrews 9:14 "*How much more shall the blood of Christ...purify your conscience.*"

The Conscience is malleable—both ways—for better or for the worse. Again. this is distinctly spiritual as the genetic code is not being overwritten.

From these quotes, it is evident that the Conscience plays a role, not only in recognizing what actions are wrong to do, but also in motivating the individual to perform altruistic deeds which have little or no benefit for the individual. The conscience being malleable, fits in very easily to the Body and Soul model for the human.

But could the conscience also fit into the Body-only model by being based in the genetic make-up of the individual? Scientism says yes. Scientism believes that there is an altruistic nature of man which is determined through the genes and is part of the evolutionary process. Altruistic acts of mankind include doing good or even risking one's life to save another whom one doesn't even know, such as others of different race, color, or creed, or even on the other side of the planet. Richard Dawkins states that origin of this altruistic evolutionary genetic selection might be as follows.[76] Early on, man's world was far smaller than what it is today. Man basically initially just interacted with kinsman, or other tribal members. Thus, because of the closeness of the individuals, any good act could have a benefit for the individual within that small group. In effect, one would hope for a reciprocal altruism. Not only that, the good deed might raise one's

[76] Dawkins, *The Blind Watchmaker*, 251-253.

status among the tribe. Thus, the genetic line of the individual would prosper because of increased likelihood of surviving and because of increased mating opportunities. In this manner, mankind, as a whole, would genetically become more thoughtful. One could say a "kind" gene was developed. This beneficial "kind" gene for those who are directly involved in a person's life grew in number as time went on.

However, man eventually spread his activities into larger and larger circles. Now, his kind activities were provided to others with no potential benefit to himself. He calls this a genetic "misfiring." As an example of a such a misfiring in the non-human animal world, he writes of the cuckoo bird who places its offspring in the nest of another species of bird. The mother bird finding this other bird in her nest feeds the bird because the mother bird's instinct is to "Look after small squawking things in your nest and drop food into their red gapes." For Dawkins, this a "misfiring" of natural selection. He then goes so far as to suggest the Good Samaritan activities of humans are similarly misfirings of the natural selection process. Perhaps, he suggests, even the human urge to adopt a child is also a "misfiring!" Dawkins clarifies that he doesn't mean this in a "pejorative" sense. It is just not what the selection process intended to do (I almost wrote "is designed to do"!). He states sexual desire for an infertile or contracepted female would also be a Darwinian mistake, for according to Darwin, sex is naturally designed for procreation. Dawkins considers this Darwinian mistake to be a blessed, precious mistake. However, Dawkins arguments are hand-waving at best.

According to Dawkins, within a few tens of 1000's of years, every human had built in altruistic genes, which can then be referred to as a conscience. Dawkins, referencing the evolutionary development of altruism, indicates that this process is just like that of genetically establishing sexual desire into the human DNA structure. As we know, the sexual drive is extremely strong. (Hopefully, the evolutionary development of a conscience is comparably continuing today. Perhaps with time on its side, the "kind" gene will get to be as strong as the sex drive gene. If this be the case, I, personally, look forward to the day when all people are genetically altruistic even though I probably will not be there to enjoy it. How altruistic of me!) As we have it now, it sometimes seems the "kind" gene must be a very minor gene with minimal influence. Hopefully, scientists will soon discover a time machine which will allow me to move into the future where with proper development everyone is kind. Alternatively, maybe we can bring back a few human specimens to mate with 21st century mankind and speed up the process of making everyone kind. Of course, maybe a "mean" gene beneficial to survival will have developed by then, in which case, no thanks!

Am I being facetious? Well, no. If one believes that behavior is controlled by molecules, what I have said could be absolutely true. But despite what I have said above, I would dread such a day. Why? Because love is so important. If biology fully controls behavior, we will no longer have free will. Without free will, one cannot choose to love. Obviously, if you believe behavior is controlled by molecules, at the current stage of evolution, biology is not doing a good job of controlling our behavior.

But that is not the case. We are in a Nature AND Nurture situation. James Fallon, PhD, is a neuroscientist who once believed all of our behavior is controlled by genetics. That changed when he had his own brain scanned, and he discovered that he should be a psychopath.[77] However, he came to recognize that, while genetics play a role, environmental factors such as family life, life experiences, personal choices can and do affect behavior. Free will plays the major role. While behavioral geneticists have spent years evaluating this, parents have known this firsthand. Having raised seven kids, I have seen the direct effects of biology on their temperament. Some children are compliant, some are aggressive, some are artistic, some are rigid, some haven't a care in the world. So, while you must love them equally, treating them all exactly the same just doesn't work.

---

[77] J. Fallon, *The Psychopath Inside: A Neuroscientist's Personal Journey into the Dark Side of the Brain* (New York, NY: Penguin Books, 2013).

So back to the issue of character traits—altruism, conscience, selfishness, etc., being an evolutionary by-product. This just is an incomplete idea. We know genes affect behavior, but that is only part of the story. If we consider the idea that a soul is conjoined with the body, we easily see that the effectiveness of the soul in regulating behavior would also be affected by the body, just as intelligence, communication, imaging, etc., are affected by the body. From a religious perspective, a grace-filled person will be more able to be a loving person. From an evidence-based approach, there are millions and millions of stories which demonstrate how an encounter with God immediately changed one's life. These stories cannot all be made up. These encounters with God cannot be encounters with molecules. Grace-filled encounters are not physical encounters like getting tackled in a football game. Instead, the grace-filled encounter is with the innermost non-physical person. It is an encounter with the "heart" of man. We say it is an encounter with the soul. Anyone who does not encounter others with one's heart would surely be called cold and rigid.

If, indeed, there is a Body and a Soul, are their other things in life which are not physical, or do not follow physical laws? If the FULLNESS OF REALITY includes true realities with no physical construct, such as love, truth, justice, friendship, kindness, and on and on, then science only defines a piece of reality—as intriguing as that piece may be. Are there other things that are part of reality that do not follow science's rules? Is there real evidence of paranormal physical events? Yes, there are, and I will present a few of these to you in the next couple of chapters. We call them Miracles.

Things to Think About

1. Are the 10 Commandments a reflection of man's conscience?

2. If the conscience is a genetic biological code, wouldn't it be subject to mutation?

3. What about love? Is it genetically controlled?

Chapter 10

Sense or Nonsense?

Are miracles real? Are they part of the FULLNESS OF REALITY that I keep talking about? The answer is–some are, some aren't. However, even if a thousand miracles were bogus, it doesn't matter. There needs to be only <u>one</u> miracle which is true, and which defies the laws of physics, to eliminate the concept that everything that happens in life comes through scientific materialism. Certainly, I have no issue with Science. I love Science! Science is utterly fascinating, but one will never have a complete picture of reality if part of Reality is left out. (I love Reality, too!) There are so many claims of miracles throughout all of history. Could they all either have fooled everyone, be outright lies, or, perhaps, with the right science, all be capable of being explained.

If miracles are real, are there some miracles ongoing and can be seen today? The answer is yes, and I have either directly seen them or have seen irrefutable evidence of them. And yes, you, too, can see the miraculous, if you so wish. Be open to the evidence. Here are a few unexplainable objects one finds in the world.

The Eucharistic Miracle of Lanciano, Italy (~700 AD-ongoing)

The most common Catholic liturgy is that of the Mass. The highlight of the Mass is the consecration of the Host, a round piece of unleavened bread, followed by the consecration of the wine. In the act of consecration, the priest will say the words Jesus Christ said at the Passover meal in the Upper Room on Holy Thursday— "This is my Body..." while holding the Host, followed by "This is my Blood..." while holding the cup (chalice) containing the wine. The Church does this because from the very first century of Christianity, the Church has always believed that when the priest says "This is my Body," the host actually becomes the body of Jesus Christ, even though the consecrated Host appears unchanged (one says that the "accidents"—the sensory elements, such as looks, taste, smell, etc., of the host remain the same) and the "accidents" of the consecrated wine, likewise, appear unchanged. However, the essence of the consecrated Host substance (the essence of what it truly is) has become the glorified body of Jesus Christ. Similarly, the Church has always believed that the substance of consecrated wine becomes the Most Precious Blood of Jesus Christ. The Church calls this consecrated Host and Wine the Eucharist.

For Catholics, the Eucharist has become the Real Presence of Jesus Christ, that is, the Body, Blood, Soul, and Divinity of Jesus Christ.[78] This is quite a claim! So, is there any physical evidence that this might be true? Physical evidence helps because physicality is such an important part of our lives, and those who believe in scientific materialism demand it. Without such evidence, they would just believe one is pulling their leg! Not surprisingly, while physical evidence would be shocking enough, this miracle is understood only through the light of faith for the most important things in life are not physical.

Naturally, over the years many individuals have had doubts that the Eucharist is really the Body and Blood of Christ. If only we had a miracle to show us the truth!

Figure 10.1: Monstrance containing Eucharist miracle of Lanciano, Italy

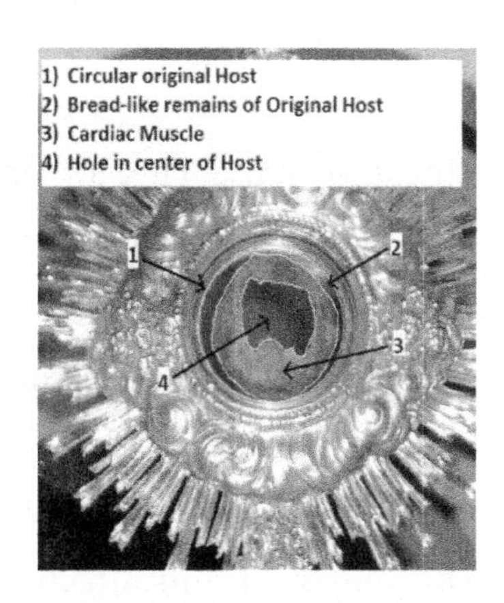

Figure 10.2: Enlargement of the image of the Lanciano host/flesh

[78]For more information, see David J. Keys, *Exploring the Belief in the Real Presence* (Bloomington, IN: iUniverse, 2016).

Fortunately, to bolster the faith of Catholics, God has provided over 300 miracles associated with the Eucharist. The most famous of these is the Eucharistic Miracle at Lanciano, Italy. In this miracle, the Host has been transformed into real cardiac muscle, and the consecrated wine has become real blood. This miracle, as you will soon see, is ongoing today.

A full accounting of this miracle can be found in numerous texts.[79] A shortened account goes as follows: Around AD 700, a Basilian monk (an order of St. Basil the Great; Basil lived from AD 329 to 379) was having severe doubts as to the reality of the bread and wine changing into the body and blood of Jesus in the Eucharist. Despite his doubt, he continued to say Mass as usual at the Church of St. Longinus in Lanciano, Italy. This all changed when one day, during a mass he was saying and just after consecrating the hosts and the wine, suddenly, the "accidents" of the bread of the Host changed into both "accidents" of bread and of flesh (which I will refer to as bread and flesh for ease in presenting the account). The very outer edge of the host remained bread, while internal and attached to the bread, flesh appeared,

---

[79] See, for instance, *Eucharistic Miracles* by Joan Cruz, or *The Eucharistic Miracles of the World* by the Institute of St. Clement I, Pope and Martyr. *Exploring the Belief in the Real Presence by David J Keys, PhD* by iUniverse. These books are listed in the "Suggested Reading" section at the end of the book.

being very thick at the outer edge and then thin near the center.

Now it is impossible for bread to turn partly into bread and flesh. And, of course, the flesh being physically attached to the bread is also impossible. There is just no scientific way for this to happen. Certainly, evolution didn't have time to do this. That would be miraculous enough, but it would be just a good story if not for the *fact* that the miraculous host and blood are still around today and have been made available to be analyzed. And analyzed, it has been.

These elements have been exposed to air and have not been chemically treated for 1,300 years, yet they still exist. Bread, tissue, wine, and blood decay with time, so the fact that these elements exist today in itself can only be termed a miracle. The flesh is not any flesh; it is cardiac muscle. No one can duplicate this. The Host/flesh remains the same as it was on the day the miracle occurred, except that the thin center portion of bread has disappeared. The blood has now coagulated into five separate pellets. In 1970, Dr. Odoardo Linoli, a professor of anatomy and pathological histology in chemistry and in clinical microscopy in Arezzo, Italy, examined the elements. The tissue and the blood within the tissue remain similar to any biopsy material obtained from an operating room, which one would send to a pathologist for examination. The pellets of blood can be liquefied and are truly human blood. In effect, when tested, the tissue and blood act as if they have just been taken from a living person.

The tissue type is striated muscular heart tissue from the myocardium (heart wall) of the left ventricle. The tissue is human and contains arteries and veins, along with the vagus nerve. The blood type is AB (about 6 percent of people have type AB blood). The blood comes from a man of Middle Eastern origin. These results were published in a peer-reviewed journal[80] and later verified in 1973 by a scientific commission appointed by the chief advisory board of the World Health Organization. All in all, their testing lasted fifteen months and included more than five hundred tests. The results were consistent with the 1970 tests. Can you imagine the shock these doctors felt when these results came in?

There is no rational explanation for the host or the coagulated pellets. Again, assuredly, no one knows how to attach cardiac muscle to a rim of unleavened bread. In all aspects, the miracle of Lanciano certainly follows the 1st century belief of the Catholic Church, continuing through today, that a change in substance really happens, that the consecrated bread and wine really do become Jesus' Body and Blood. Interesting is the cardiac muscle. Catholicism has a strong devotion to the heart of Jesus, and for this miracle to be a 1300-year-old heart muscle of a Middle Eastern man is certainly faith edifying.

---

[80] *The Sclavo Notebooks in Diagnostics* Collection #3 (1971).

In 2014, I was fortunate enough to go to Lanciano and was able to stand within three feet of the miraculous Host. You can do that, too. Prove it to yourself. Go, stand within three feet of this ongoing miracle, and see this with your very eyes. If you think, someday, someone will duplicate this "miracle" through science. All I can say is, "Don't hold your breath!"

You are likely wondering whether these types of miracles still occur. Pope Francis was involved in a Eucharistic miracle in 1996 while still an auxiliary Bishop in Buenos Aires, Argentina. In 1992, a Eucharistic miracle occurred at Santa Maria y Caballito Almago Catholic Church. The evidence was stored away. In 1996, a second Eucharistic miracle occurred at the same Church. This time, Auxiliary Bishop Jorge Bergoglio (the future Pope Francis) was involved. The samples were stored away. By 1999, the samples had not decayed, and Bergoglio had the tissue evaluated by a local specialist and also sent to the University of Sydney in Sydney, Australia, and to specialists in San Francisco. The nature of the sample was not told to the investigators. And, guess what, the Host had turned into cardiac muscle from the same region of the heart as had happened at Lanciano. The blood cells were intact, and the blood is Type AB from a Middle Eastern male. In effect, they concluded, the tissue was alive.

Six years later, in 2005, a sample was sent to internationally known forensic medicine specialist Dr. Frederic

Zugibe. Again, he did not know the nature of the sample. Upon examination of the sample, Dr. Zugibe saw living blood cells moving about in the slide samples. He noted an accumulation of white blood cells, indicating that the person from whom these samples were obtained, had been traumatized. The sample he noted was from the heart near the left ventricle. Can you imagine the shock he felt when he discovered the origin of the sample? Zugibe continued with his investigation and was then able to compare the tissue sample to the sample obtained in Lanciano, Italy. He concluded they were from the same individual.

What to make of this? None of this is possible according to modern science. Yet, it exists. These truly are miracles. Since these miracles strengthen the belief in the doctrine of the Real Presence, I simply call them Doctrinal miracles, with no other purpose than to strengthen our faith. What do atheistic scientists call them? "Ignored."

The Shroud of Turin

Another physical evidence of a miracle that can be viewed is the Shroud of Turin. The Shroud of Turin is proposed by many to be the burial cloth of Jesus from circa 33 CE. The cloth shows the image of a man that was crucified. The image on the Shroud cannot be reproduced even today, despite years of scientific attempts, using every piece of technology modern man has at his disposal. Many scientists have labeled it as an example of an Acheiropoieta. Acheiropoieta are works of art, religious in nature, which have not been made by human hands. This would, of course, be miraculous if that is to be the case.

Before I delve into the reasons it is termed an Acheiropoieta, let me, first of all, give you some facts. The Shroud is

Figure 10.3 The Shroud of Turin on Display in 1931, in Turin, Italy (Picture taken from the Exposition in Turin in honor of the wedding of Prince of Piedmont, Umberto of Savoy, to Princess Maria Jose` of Belgium).

about 436 cm Long (14'4" long), 110 cm wide (3'7" wide). The shroud is made with a herringbone weave, indicating it was very expensive. On the cloth is a frontal and dorsal image of a naked man who apparently was crucified as determined by the markings on the cloth.

The cloth is so long because the cloth was laid lengthwise on the burial stone and the man's feet was positioned at one end of the cloth and the man's head placed near the middle of the cloth. The cloth was then pulled over the man's head and brought down to his feet. The man himself was about 5'11' and about 160 to 175 lbs. Assuming for now that the shroud is 2000 years old and originated out of Palestine, the man would have towered over the average man of his day, since the average height at that time was around 5'4". In other words, by stature he would have been a natural born leader. None of this, of course, gives any reason to consider the image to be "not made by human hands." It is the image itself which is so uncanny and irreproducible.

The shroud image is in two colors–red and a yellowish-brown straw color. The red arises out of the bloody areas. Yes, there is true human blood on the shroud. Could this blood have been painted on a flax linen? Well, no! First of all, there is no evidence of any brush stroke. Also, it is certainly not painted since there are no paint pigments present. That takes care of the paint idea. Again, the red markings are actually blood! Could an artist have used real blood? No. There is so much detail which no one knew existed prior to the 20th century. No one looking at the shroud prior to the 20th century could have detected all the intricacies in the image without specialized modern day equipment.

The blood itself is strange. It is red. Blood exposed to air does not stay red. It turns a dark brown. The blood itself has been typed as AB, although the classification could be due to the effect of age on blood samples. The blood is known to be from a Jewish man. Not only is blood found, but there are areas of venous blood flow and arterial blood flow. Human serum albumin and clot retraction rings are also found on both the front and back of the image, which naturally would have been included with real blood. No one really knew much about venous versus arterial, or human serum, or clot retraction rings until the 20th century. The blood necessarily somewhat soaked through the cloth. The blood patterns occur at areas of trauma, plus areas where the blood would have flowed had the victim have been vertical (as on a cross), or horizontal (as having been carried to a grave site). What's more, the actual image on the cloth is the mirror image of the real image. Right is left and left is right. This would occur if the cloth were placed on the skin. In order to see the image, one would have to turn the cloth over. In that process, right would become left, etc. You can easily try this out for yourself by putting a cloth on you and put a safety pin on the right side. When you flip it over the safety pin will be on the left. These characteristics pretty much guarantee that the image is not a painting, but rather is an image on a cloth that was truly wrapped around a crucified body. There is even more evidence that the cloth was on a real body, but I will refer you to Mark Antonacci's book *Test the Shroud*, if you would like to go deeper.[81]

So, one has a stained cloth with real blood which was on a real body, but that doesn't make it a 2000-year-old cloth with an image "not made by human hands." That's right.

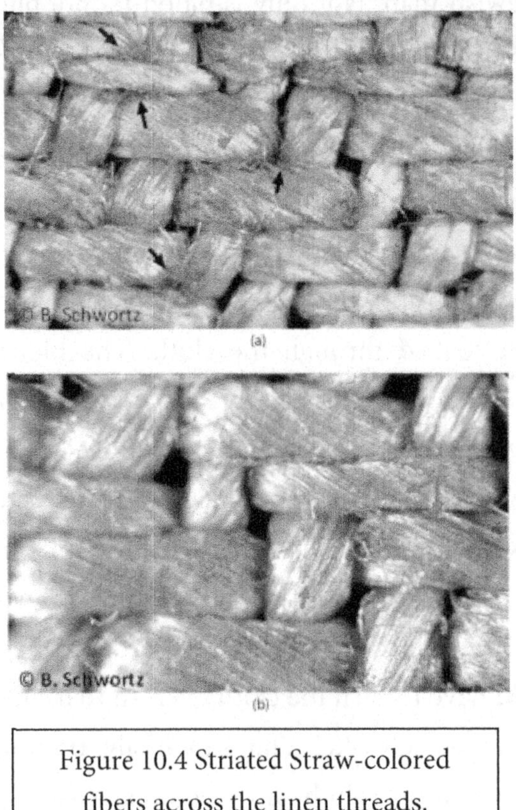

Figure 10.4 Striated Straw-colored fibers across the linen threads.

[81] Mark Antonacci, *Test the Shroud: At the Atomic and Molecular Levels* (Forefront Pub Co., 2016).

Amazing as that is, there is more. As you may guess, I'm not done yet. That the cloth was on a real body is the easy part! Let's look at the straw-colored images. First of all, there are no straw-colored images under any of the blood. However, the image was made, it is evident that the blood was there first and that the imaging process couldn't penetrate the blood. There is no paint pigment on the shroud, so that is not a reason for the color. Instead, the image is the effect of the oxidation of the fibrils of the threads of the linen. All fibrils oxidized have the same shade of straw-color. Evidently something happened to the threads that caused the oxidation. Only the fibrils above and below the body are oxidized whether are not they were actually touching the body. But not every fibril above or below the body is oxidized, only the most superficial fibrils on top of the thread within 20 microns (0.00079 inches) of the surface were oxidized. This means the oxidizing agent did not penetrate much at all. But not all superficial fibrils within 20 microns oxidized. One may be oxidized, yet the one next to has no sign of oxidation. One would say the image is striated. As I said above, the color of the oxidized fibrils all have the same shade. The reason some appear darker is because there is a greater density of oxidized fibrils there. See Figure 10.4.

You might be thinking that there was some chemical on the man's body which, when it came into contact with the cloth, caused the oxidation. Well, that is not true either. Some areas would not be touching the body, for example, areas of the neck near the chin. There would be a draping, but no contact. However, even though those areas did not contact the body show detail from the body. These areas are lighter (fewer fibrils) compared to an area like the chest where the shroud was actually on the skin. This is also clearly seen when one compares the right and left thigh of the man. Apparently, the left knee was flexed, so the left thigh was up and the right down. The raised thigh appears much darker than the lowered thigh because there the shroud touched the skin. Scientists have found that since the density of oxidized fibrils varies with projected distance from the body, getting lighter as one has less direct contact with the skin, that the image contains 3-dimensional information. Thus, the dimensions of the body can be estimated.

Wow! If this image were created by an artist, that person would have to have been extremely clever. But could the image have been made due to some gas emanating from the body, such as ammonia gas? Well, no, because of another unexplainable feature of the shroud. Studies on the shroud show remarkable detail even for parts of the image which are not touching the skin. Normally, the image would be expected to get fuzzier as it gets further away. The image on the shroud does not show this. In fact, it has been stated that it appears "as if" laser beams were pointed vertically from the body, striking the cloth. Note, I said "as if," not that laser beams were used. It appears that either the source of imaging was projected outward vertically from the body or that the cloth fell vertically through the body.

I would say this is all gibberish, if, in fact, it wasn't true. Independent of how old the cloth is and whose image is on the cloth, this is an image which science cannot begin to replicate today, much less at any time in the past. One thing we do know with a reasonable confidence is that the image either is, or was created to be an image of Jesus Christ. How can I say this? One need only look at the Gospels in the New Testament to see the description of Jesus' Passion and Death to see that the shroud shows everything stated.

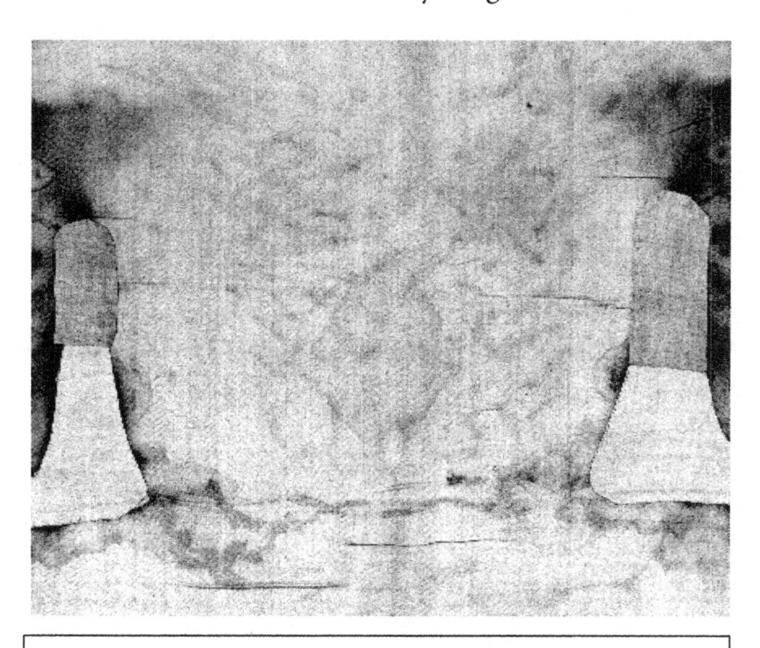

Figure 10.5 Posterior view of the Scourge marks across the back of the man in the Shroud of Turin.

- Jesus was a young man of stature in the prime of his life. The shroud shows a man of stature—a muscular 5'11 vs a typical 5'4" average Jew. Conservative odds of that happening are 1 in 20.
- Jesus was beaten about the face. The shroud image shows this. This probably was not uncommon. Conservative odds are 1 in 1.
- Jesus had a crown of thorns put on His head. The shroud shows multiple blood marks consistent to what one would see if thorns were stuck in His head. This would be extremely uncommon. In fact, Jesus is the only person in history who was known to have a

Figure 10.6 Typical Roman Flagrum used for Scourging.

crown of thorns placed on His head and was crucified. Conservative odds are 1 in 5000.

- Jesus was scourged. The shroud shows anywhere from 100 to 120 scourge marks across the chest, back, and legs. It was common for the Romans to use a three-barbed flagrum. The maximum number of lashes by Jewish law was 40. Typically, those intended to be crucified had fewer. However, Pontius Pilate thought maybe the people would choose Jesus for release if He had been fully scourged. Conservative odds are 1 in 3.
- Jesus fell while carrying the cross. The shroud shows knee abrasions and a dislocated shoulder. Conservative odds are 2 in 3.
- Jesus was nailed to the cross. The shroud shows this. Nailing was common, but so was tying the arms to the crossbeam with a rope. Conservative odds are 1 in 2.
- Jesus had a lance put into His side to prove He was dead. The shroud shows this. This was not common. Usually, the soldiers would break the legs to ensure the person on the cross would soon die by asphyxiation (could no longer raise up to breathe). Conservative odds are 1 in 10.
- Jesus was a Jew. From analysis of the blood on the shroud, the man in the cloth was Jewish and not

someone from some other part of the Roman Empire. Conservative odds are 1 in 5.

- Jesus was buried in an expensive linen and laid in a tomb. The shroud with its herringbone pattern would have been an expensive linen. It is more than reasonable that the rich disciple and Sanhedrin member, Joseph of Arimathea, would have purchased an expensive cloth. In addition, it would be rare for a man whom the Romans crucified to be placed in any expensive linen, and certainly not in a tomb. Those crucified were placed in a common grave and sometimes were eaten by the dogs. Conservative odds are 1 in 100.
- Jesus was raised from the dead fewer than 40 hours after he died. Therefore, he would not have undergone putrefaction. The shroud shows no evidence that the body it contained ever decayed. Conservative odds are 1 in 500.

The probability that someone other than Jesus was intended to be in the image is 1/20 x 1/1 x1/5000 x 1/3 x 2/3 x 1/2 x 1/10 x 1/5 x 1/100 x 1/500, or 1 in 2.5 trillion. As you can see, by this simple analysis, it is unreasonable to think anyone other than Jesus was intended to be the image on the cloth.

So far, it is apparent that the Shroud is miraculous in that the image is not reproducible, even using our most advanced equipment to generate the equipment. Also, it is reasonable to say the image on the Shroud is intended to be Jesus of Nazareth. But is it the right age to possibly be the actual burial cloth of Jesus of Nazareth?

It is sad to say, but the experts who had done all the preliminary work on the Shroud were not allowed to take part in the Carbon 14 dating. A group headed up by the British Museum in 1988 selected three sites (Oxford University, Zurich Polytechnik, and the University of Arizona) to perform the Carbon 14 dating. Unfortunately, the samples for testing were taken from the upper left corner of the Shroud—a place which had been repaired. This introduced cotton fibers into the samples when the Shroud's fibers were strictly flax. None of the three sites tested the samples for contamination of their samples as was part of the protocol. As the cotton fibers are younger than the flax, the test results came out younger, and what is more, they announced their results publicly in a news conference on October 13, 1988, that dated the Shroud from 1260 to 1390 CE, indicating the Shroud is a fraud (as a possible burial cloth of Jesus).

Also, the data for the three sites showed a wide variation, which should not happen for a uniformly aged material. This was not investigated. Apparently, those dating the Shroud were unaware that there were drawings of the Shroud, even showing burn marks, the herringbone pattern, and other definitive indications of the Shroud, from 1192—outside the measured range.

Years later on January 20, 2005, a peer reviewed scientific paper by Raymond N. Rogers, retired Fellow of the Los Alamos National Laboratory, and one of the original researchers in STURP published—*"Studies on the radiocarbon sample from the Shroud of Turin."* The paper concludes, *"As* unlikely as it seems, the sample used to test the age of the Shroud of Turin in 1988 was taken from a rewoven area of the Shroud." Whether the amount of contamination is sufficient to make the Shroud appear significantly younger than it the original Carbon 14 dating is debatable.

However, there are other studies of the linen which suggest a much older date. Ancient records indicate that there was in 544 CE an extraordinary image "not made by hand" which was preserved in Edessa (modern Urfa, Turkey). The large cloth was folded in such a way that only the face was shown. This cloth was transferred to Constantinople in 904 CE. Eventually the cloth was unfurled and was determined to be a burial shroud. Furthermore, a written homily from that time which mentions the full body image on the Shroud has been discovered.

There have also been studies done on linens known to be from the 1st century. A very comparable herringbone linen has been found at Masada, the site of the last stand of Israel before the Romans completely destroyed the Jewish homeland. The flax and weaving techniques of the Shroud are also comparable to known techniques from the 1st century. Such dating by looking at comparable linens places the Shroud from Palestine in the 1st century.

Further studies have shown that pollens specific to Palestine have been found on the Shroud as have limestone dirt particles, which again are specific to Palestine. Everything keeps pointing to the Shroud's being a 2000-year-old Palestinian linen.

Let's go back to the image formation process. As I have stated, no technique has been developed to duplicate the Shroud imaging, which doesn't mean they haven't made progress. There has recently been a technique developed, which by using proton radiation, has been able to produce straw-colored fibers in linen.[82] Some of the fibers also show as striated. At this point, no one is saying they can produce a human figure with the proton radiation patterns. And, of course, where would the protons come from in the first place? Well, the body is full of protons and neutrons in the various atoms making up the molecules which form our bodily elements. If indeed, as Christian teaching states, Jesus' body when resurrected became a glorified body, perhaps there was a spontaneous disintegration of Jesus' crucified body, releasing both protons and neutrons (and electrons), followed by the creation of His new body. Sounds farfetched,

---

[82] Arthur Lind, *Image Formation by Protons* (International Conference on the Shroud of Turin: Pasco, WA July 19-22, 2017).

but nothing else even comes close, so hang in there. The low energy protons then would have irradiated the cloth producing the straw-colored fibrils. (It would take some months for the oxidation to occur to make the image viewable.) But the concurrent release of neutrons would have another effect of interest. The neutron radiation would have increased the number of Carbon 14 time of the cloth, making the cloth appear much younger when Carbon-14 dating would later be applied. (Having the sample contaminated with cotton fibers is not necessarily the only cause for the younger dating of the Shroud.) Thus, one has a theory of image formation which also is a solution to the Carbon-14 dating issue.

Finally, whether this theory of image formation comes to fruition, time will only tell. In any respect, this discussion on the Shroud of Turin has demonstrated that an image not made according to any known laws of physics, biology, chemistry, or artistry does exist. And Science cannot explain it, except, that is, by appealing to the spirituality of the Resurrection of Christ. Particularly, since this image strongly suggests the "reality" of the Resurrection of Jesus Christ, I would call this a Doctrinal miracle.

Summary

What we just reviewed are two instances of material objects that cannot be explained by science. One does not have to believe that the Eucharistic miracles are actually the body and blood of Jesus Christ. However, one does have to acknowledge that there is no possible scientific way to explain how cardiac muscle tissue could be attached to bread. And certainly, how could such cardiac muscle exist for 1300 years without decay? The fact that it matches Church doctrine of the Eucharist is probably another area one needs to investigate.

Also, consider that one does not have to believe that the Shroud of Turin is truly the burial cloth of Jesus Christ to acknowledge that the image on the Shroud of Turin is not creatable by science. Whether there was some 14th century individual who created this image or not, the image itself is miraculous. It is beyond the realm of Nature. One can't just ignore these physical facts. One must then consider whether there is something outside everyday science that is capable of creating such physical masterpieces. Understanding there is a God fits the actual data quite well. Now, let's look at some other miraculous events.

Things to Think About

1. What are, for you, the most intriguing aspects of the Eucharistic miracle at Lanciano?

2. Regardless of whether the Shroud is the burial cloth of Jesus, what are for you the most intriguing aspects of the Shroud of Turin?

Chapter 11

A Cloak for All Seasons?

In chapter 10, we looked at two completely irreproducible, scientifically unexplainable objects—the Eucharistic Miracle at Lanciano, Italy, and the Shroud of Turin. The Eucharistic miracle is certainly not a work of art. It is also very difficult to call the Shroud of Turin a work of art, even though it does depict an image of a man. Certainly, every image created is not art. One simply has to look at the many

Figure 11.1 Image of Our Lady of Guadalupe

selfie images taken with today's smartphones. What we are going to look at next is an image created in the 16th century, which is certainly not painted art, but is more, so to speak, like a polaroid (if you can remember what that means) with many artistic aspects. But you might say, "Polaroid-like products were not available in the 16th century." That is so true, and that is the point! The capture of the image is but one of the aspects of this image that makes it so miraculous! The image is found on the cloak (Tilma) of St. Juan Diego. The image is that of Mary, the mother of Jesus, who, in this representation is called Our Lady of Guadalupe. As you will see, the image left behind is another example of Acheiropoieta—an image "not made by human hands." For those interested and able, you can see this image every day of the week at the Basilica of Our Lady of Guadalupe in Mexico City.

Now to fully understand the image of Our Lady of Guadalupe on the Tilma of St Juan Diego, we must first look at the historical aspects of life in Mexico in the 16th century. The account begins with the arrival of Cortez and the Spanish Conquistadors in 1519. By 1521, Cortez had conquered the Aztecs, and Spain ruled the land. Cortez and his men looked down upon the Aztecs because they believed that the Aztecs were children of the devil and, therefore, subhuman. Why did they believe this? They believed this because the Aztecs believed that they were children of Topiltzin Quetzalcoatl—a feathered Serpent God. Those raised in a Judaeo-Christian culture would, of course, based on the book of Genesis, associate one's being a child of a serpent as being a child of Satan. In addition, the Aztecs practiced human sacrifice, including child sacrifice, and practiced cannibalism, eating the bodies of those sacrificed to the gods. This further suggested to the Spaniards that the Aztecs aligned with the devil.

By 1524, the Franciscan missionaries came and had a different view. They saw the Aztecs as human beings, like the Spanish, but needing education and evangelization. They treated them accordingly as men and women made in the image and likeness of God and began to baptize some of the Aztecs. The Franciscans were met with hostility from both the Spaniards in control and from the Aztecs, who, of course, did not like the Spanish, who were mistreating their people.

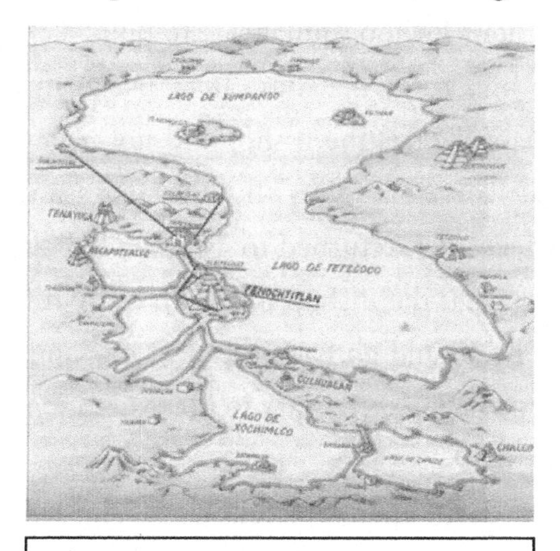

Figure 11.2 Map of Area near Tepeyac Hill

Among those who converted was a man, Juan Diego, and his wife, Maria Lucia, both of whom were baptized in 1524.

In 1529, Maria died, and Juan Diego moved from his home at Cuautitlán to live with his Uncle Juan Bernadino in Tulpetlac. Juan Diego was a pious man and would walk 13 miles from his uncle's house to the Church of Saint James in Tlatelolco for his weekly catechetical instruction. On Saturday, December 9, 1531, after 9 miles of walking, as Juan Diego passed by Tepeyac Hill, the former site of a pre-Columbian temple to the mother goddess Tonantzin, he heard birds singing and the Earth responding in song. Juan Diego then saw a woman clothed in a radiant light. She called him her "Beloved Little Son" and identified herself as the Virgin Mary, the Mother of the One True God, Giver of Life, Lord of Heaven and Earth, Creator of Man. She asked Juan Diego to request Bishop Zumárraga to build a chapel at the foot of the hill. Later that day, Juan Diego met with the Bishop, but, of course, the Bishop did not believe him and asked that he come back later. In other words, he blew him off. Juan Diego then returned to Tepeyac Hill and told the Virgin Mary what the Bishop had said. Mary responded by telling him he should go back the next day and repeat the request. Juan Diego asked Mary to no avail to send someone greater than he.

The next day, Sunday, December 10, 1531, Juan Diego again met with the Bishop. The Bishop really didn't trust Juan Diego and told him to ask Mary for a sign before he would do anything. Furthermore, the Bishop had two men follow Juan Diego as he headed back to Tepeyac hill. Amazingly, the two men lost sight of Juan Diego when he got to Tepeyac Hill. This was particularly amazing since Tepeyac Hill is a barren, almost lifeless piece of ground. There was no real good place for Juan Diego to hide. Mary, then, met with Juan Diego and told him to return the next day and she would give him a sign.

But things did not go as planned. The next morning, Monday December 11, 1513, Juan Diego's uncle, Juan Bernardino, became deathly ill. Oh, what to do? Obey the Mother of God, or help his uncle out? Juan Diego elected not to return to Tepeyac Hill but, instead, reverted back to his Aztec upbringing and sought a shaman to heal his uncle.

Well, that didn't work, so on the next day Tuesday, December 12, 1531, Juan Diego decided to get a priest to administer the last rites to his uncle. There was just one major problem; to get to a priest in Tenochtitlan, Juan Diego had to walk past Tepeyac Hill. Juan Diego felt that he was in trouble for disobeying the Mother of God! Imagine that. So, Juan Diego chose a different, longer path around Tepeyac Hill, so that he might avoid Mary while on his way to find a priest for his uncle. Well, that didn't work. Mary saw Juan Diego and called him to her. Juan Diego told Mary his plight, but she responded by telling him not to worry and that she had already visited his uncle, and he had been healed. Then she revealed herself as the Perfect Virgin, Our Lady of

Guadalupe. Next, Mary told Juan Diego to go to the top of the rocky, barren Tepeyac Hill and bring back flowers. (There were no flowers there on Sunday. Flowers do not naturally grow there at that time of year.) Juan Diego went as requested, and to his surprise found flowers, including Spanish Castilian roses. Now, this was December and there would be no flowers there, much less Spanish Castilian roses. He brought them down to Mary. Mary arranged the flowers in Juan Diego's tilma and sent Juan Diego on his way to see the Bishop. (An important point needs to be made here. The tilma, a predecessor to the poncho, was worn by the Native Americans in the front like an apron and was made of Cactus fibers—fibers which are stripped from Cactus leaves much like fibers which are stripped from celery or other such plants, only the fibers are larger and coarser, being about 1 mm in diameter. Coats made this way were sufficient for

Figure 11.3 Typical Aztec-style of Painting

daily use but did not last long—maybe 10 to at most 20 years, if lucky.)

When Juan Diego arrived at the Bishop's residence carrying the roses concealed in his tilma, the guards would not let him in. When one of the guards forcibly reached into the tilma and tried to grab the roses, the roses simply vanished. With all the commotion, others soon gathered at the entrance. At that point, Juan Diego and the others were allowed to enter into the Bishop's office. Juan Diego told Bishop Zumárraga that he had the sign from Our Lady that he had asked for. He lowered his tilma expecting the flowers to fall on the floor, but instead a flash of light occurred and there on his tilma was an image of Our Lady of Guadalupe! Within two weeks, a small chapel was constructed at the base of Tepeyac Hill.

Concerns

What of the image itself? Wasn't this just a man-made terrific painting, intended to draw the Natives into the Catholic Faith? Well, one thing is for sure, the image certainly isn't Aztec art. Aztec art was much more two-dimensional. (See Figure 11.3) In comparison, the artistic quality of the Guadalupe image is considered to be quite remarkable, almost a breath-taking image of Our Lady. So, it is quite unlikely that an Aztec artist painted it. But what about the Spanish? They had some great artists. In fact, there is Spanish art, which resembles certain aspects of Our Lady of Guadalupe's image. Side by side in Figure 11.4 is the image of Our Lady of Guadalupe and a painting of Our Lady of Mercy by Bonant Zaortiga (1403-1446). (Zaortiga's painting can be found at the shrine of the Virgen de la Carrasca de Blancas (Teruel)). Note, the tilt of the head, eyes looking down, the hand position, the brooch, the gold trim around the cape, and the fur trim at the wrists. Hmmmm!

Another point, looking at the painting of Our Lady of Mercy, you will see that Our Lady of Mercy has a crown and Our Lady of Guadalupe does not. Apparently, there were those who thought they could improve on the design of the image, so someone added a crown to the image when they created a copy which was sent back to Spain and eventually used as a standard for the Christian fleet when they fought the Turks at the Battle of Lepanto in 1571.

Figure 11.4 The Image of Our Lady of Guadalupe, the Painting of Our Lady of Mercy, a later painted copy of Our Lady of Guadalupe, and a sculpture of Our Lady at the Monastery of Guadalupe of Caceres.

Here is another Spanish work of art for comparison's sake. The comparison image is a sculpture of Our Lady, placed in front of the choir at the monastery of Guadalupe of Caceres, Extremadura, Spain in 1499. In this case, note the crescent moon, the angel holding up Mary, the stars on the dress, the sun pattern about Our Lady. This is another Hmmmm! Furthermore, let's not forget the name given by the Virgin Mary to Juan Diego—Our Lady of Guadalupe. To top it off there was a Native American who was a Native American artist trained by the Spanish with the name of Marcos Cipac de Aquino (? - 1572). Some identify de Aquino as the artist of the tilma.

At this point the authenticity of the image as being "not made by human hands" and whether an apparition really occurred or whether the event was contrived by the Church or someone else might come into your mind. Another issue is that the document *Nican Mopohua* written in the Nahuatl language by Antonio Valeriano describing the events of the apparition of Our Lady of Guadalupe was not published until 1649—over 100 years after the event, when it was included in a book entitled *Huei Tlamahuicoltica* (The Great Happening) written by Laso de la Vega. (Antonio Valeriano was a Native American scholar educated by at the Franciscan Santa Cruz College in Tlatelolco, and knew Juan Diego personally.) However, some scholars believed that Laso de la Vega created the document himself. This was all to put to rest in 1995 when the original document *Nican Mopohua* and a death certificate of Juan Diego were found in the New York Public Library![83] It turns out that when the Americans occupied Mexico City in 1847 at the conclusion of the Mexican-American war, the American commanding officer Major General Winfield Scott sent a collection of valuable documents back to the United States. Eventually, these were filed away at the New York Public Library. (Sounds like the Ark of the Covenant being stored away at the Smithsonian Institute in the movie *Raiders of the Lost Ark.*) The document *Nican Mopohua* was discovered in 1995 by Fr. Xavier Escalada, S.J., who brought it to public attention and published it in 1997. These documents, plus many other documents concerning the life of Juan Diego, along with an approved miracle associated with the intercession of Juan Diego (a completely different topic to discuss), allowed Juan Diego to be canonized by Pope John Paul II on July 31, 2002, in Mexico City.

So, at this point, there is at least documentation that Juan Diego existed, but what is miraculous about the apparition? Is there sufficient reason for one to give up their materialistic view of life and recognize that there are events which the laws of physics cannot explain?

---

[83] Górny Grzegorz, et al. *Guadalupe Mysteries: Deciphering the Code* (San Francisco: Ignatius Press, 2016), 24.

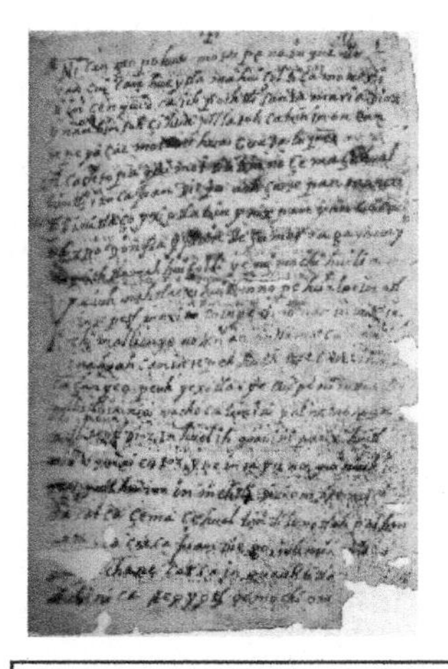

Figure 11.5 The *Nican Mopohua* and an image contained wthin the Death Certificate of Juan Diego.

The Evidence

I think we can agree that based solely on what I have just written concerning the account of Our Lady of Guadalupe, one would have to say, "Nice image, nice story, but there is no overwhelming evidence that it an apparition really occurred or, at this point in the account, that the image is not man-made." and I would agree. Ah, but there is additional substantial evidence that it is true—evidence that has remained almost 500 years after the event, evidence which defies scientific explanation, physical evidence that a miraculous event had occurred—the tilma itself.

1. Let's begin by first stating that it is unheard of that a tilma from the 16th century would still be around, much less one that is in great shape. Indeed, it is in great shape in spite of the fact that it has been touched by probably hundreds of thousands of people over the years. It has had thousands and thousands of candles lit before it. In 1753, Michael Cabrera, a well-known Mexican artist, saw pilgrims touch the tilma with over 500 different objects within two hours! The tilma was never temperature and humidity controlled for much of its existence and certainly was not hermetically preserved. Over time, various artists have tried to duplicate the image on cactus fiber material but have always failed. In fact, in 1789, Dr. José Ignacio Bartolache, a noted mathematician, tried an experiment. He commissioned eight artists to paint eight copies of the image of Our Lady of Guadalupe onto tilmas which were similar to Juan Diego's. All of the tilmas deteriorated within 8 years in the Mexican climate and had to be discarded.[84]
2. The tilma is not just a single cloth, rather it is composed of two separate cloths, which are attached together by a vertical, fine agave thread, not by a tight, strong thread, at least as strong as the cactus

---

[84] Grzegorz, et al., *Guadalupe Mysteries*, 24.

fiber threads. The two sections are joined together by a loose tacking. Amazingly after all these years of hanging, the two cloths have never separated.

3. The image itself is somehow embedded into the cactus. There is no sizing that was applied. The image of Our Lady of Guadalupe is extremely involved, but there is no preliminary drawing under the image as

Figure 11.6 A modern day Agave plant.

examined by using infrared cameras. One would normally expect that a preliminary drawing would have been made. Furthermore, the image is similar to the Shroud of Turin as previous discussed in chapter 10 in that there are no brush strokes. In fact, there is no paint over nearly all of the image (someone did add some paint in a few areas, but this is easy to see). The fibers are simply internally colored as determined by Richard Kuhn, PhD, in 1936.[85] (He won the Nobel Prize in Chemistry for other work two years later.) Since there are no brush strokes, the image is much more like a Polaroid or a digital camera image in that the image was captured all at once, but, of course, no such technology like that existed in the 16th century. No one, not even 21st century artists, can duplicate the actual imaging process. This one characteristic is enough to make the image an Acheiropoieta—"not made by human hands"—in other words, a miraculous image.

4. Furthermore, similar to the Shroud's image, which changes color as one moves closer or farther away, the tilma's image shows Our Lady of Guadalupe's skin to be olive-colored at a distance and light-colored at closer distances. This process, called iridescence is not achievable by hand painting, but does occur in nature as with certain butterflies.
5. The colors on the tilma have remained vibrant, not faded at all as one would expect of a 500-year image exposed to the light. As mentioned, burning candles were placed before the Tilma for at least a few hundred years.

---

[85] Grzegorz, et al. *Guadalupe Mysteries: Deciphering the Code*, 266.

6. As I mentioned previously, the cactus fibers are about 1 mm in diameter. The weave is not densely woven and is very uneven. According to a study performed by individuals from the Kodak corporation, the area of the tilma which has the image upon it is soft, like silk. The areas where there is no image remain very coarse. The backside of the tilma is all coarse. In addition, there is no sign of varnish to protect the image. Remarkable!
7. The unevenness of the weaves actually adds depth to the image, particularly the face.

At this point, the mere fact that the tilma and its image still exists is a miracle, but the impossibilities continue.

The Evidence Gets More Involved

8. Both eyes of Our Lady of Guadalupe have images of people when viewed under magnification. According to the account, a flash of light occurred as the tilma was opened up before the Bishop. Well, a flash of light can cause reflections on the outer surface of the cornea of the eye. Such reflections are known as Purkinje-1 images. These images can be captured by a camera. See Figure 11.7 (a), (b), and (c) below.

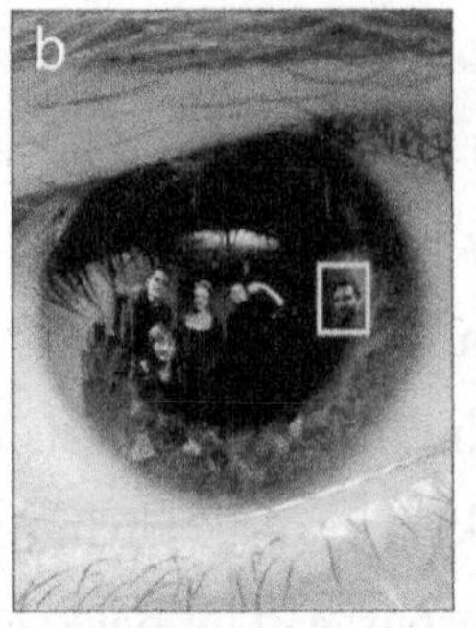

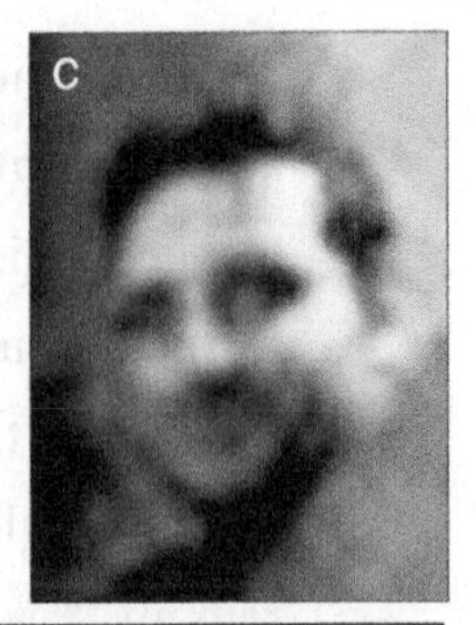

Figure 11.7 (a) An image of a young man obtained as a result of a camera flash. (b) An enlarged image of the eye shows the reflections of multiple individuals on the surface of the cornea. (c) The image of one individual in the image has been further enlarged for greater clarity.

[86]

Similarly, it has been noted that as one increases the magnification of the Guadalupe image and then adds digital image processing, images of persons within the room are seen in the eyes. You may think that such resolution is not possible on the threads of the tilma. However, the image is said to reside as on a thin film above the threads. How bizarre. (See Figure 11.8) Furthermore, the individuals in the image of the right are positioned slightly differently from the images of the same persons in the left eye, as expected due to the different vantage point of the right versus the left eye. Also, the images of Juan Diego and Bishop Zumárraga

[86] Jenkins R, Kerr C (2013) Identifiable Images of Bystanders Extracted from Corneal Reflections. PLoS ONE 8(12): e83325. https://doi.org/10.1371/journal.pone.0083325

have been compared to and correspond to images of the same men made in portrait paintings.

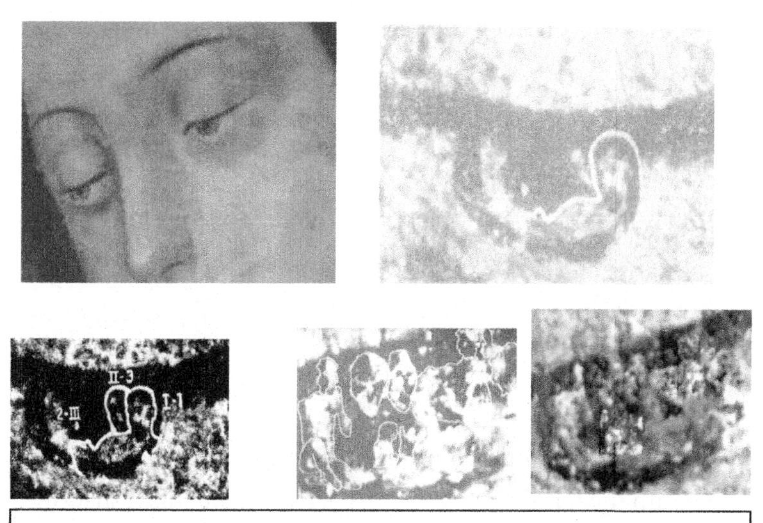

Figure 11.8 Images seen within the left eye of Our Lady of Guadalupe, as they are blown-up and computer enhanced.

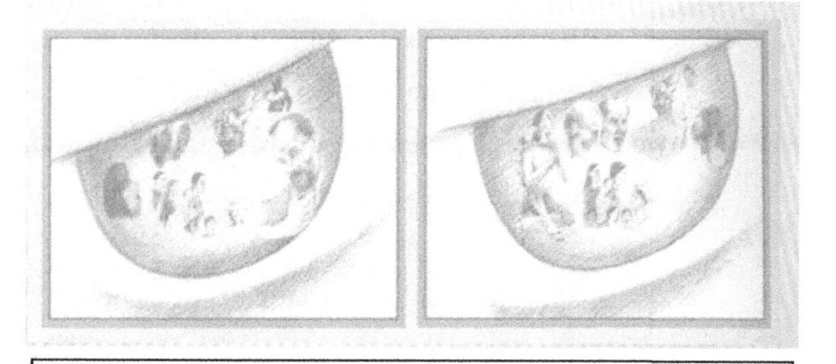

Figure 11.9 Further Processing of both eyes, showing slight shifts in the positioning of the figures due to the separation of the eyes. An artistic rendering is shown.

Figure 11.10 Additional processing of the image of the Left eye with the family image removed, leaving behind a shadow of the Our Lady of Guadalupe.

9. The tilma has been indestructible. I have already discussed the fact that natural causes should have destroyed the tilma at least 450 years ago. However, other agents should also have destroyed it.

   a. In 1789, nitric acid was accidently spilled on the Tilma while trying to clean the frame. This should have burned a hole in the fabric. The person, who spilled the acid tried to clean off the acid by rubbing the picture with a towel. This should have further damaged the image. No long-term effects were seen on the part containing the image.[87] On the non-image part, water marks can be seen (from cleaning up the nitric acid).

b. In 1921, political dissidents tried to blow-up the tilma. A bomb in a vase hidden among flowers was placed before the tilma. The bomb exploded shattering windows in nearby buildings. A metal crucifix positioned next to the tilma became bent over due to the force of the bomb. The tilma and the plain, ordinary plate glass in front of the tilma was completely undamaged.

10. Examination of the pattern of the 46 stars on the tilma has been shown to match the pattern of stars in the heavens on December 12, 1521, at 6:00 AM. However, the match of the star pattern is only seen when viewing the tilma from the Sun's position (heliocentric), not from the view of standing on the earth (geocentric).
11. The tilma has a number of flowers depicted on Mary's dress. It turns out that the flower positions correspond when scaled properly to the locations of the various major volcanoes in Mexico. What a coincidence!

---

[87] Grzegorz, et al. *Guadalupe Mysteries: Deciphering the Code,* 261.

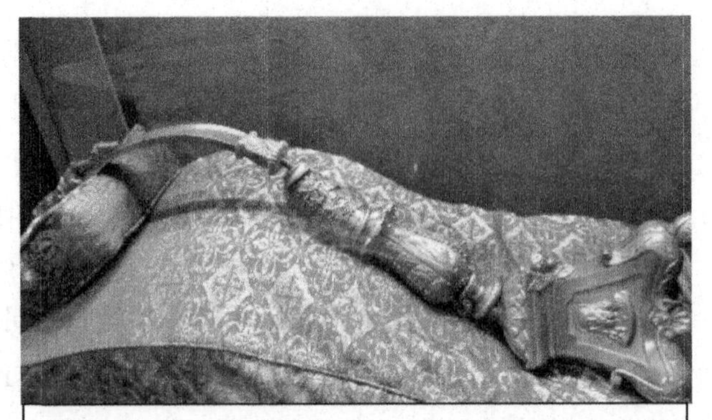

Figure 11.11 The metal cross which stood next to the Image of Our Lady of Guadalupe as it looked after the explosion in 1921.

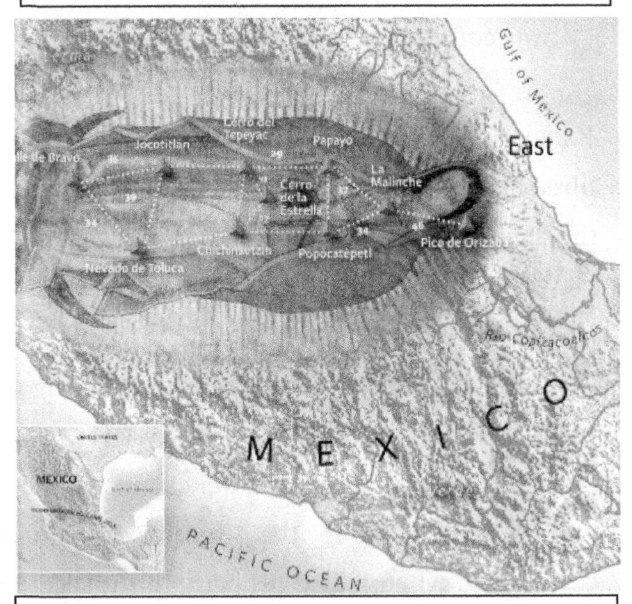

Figure 11.12 The Image of Our Lady of Guadalupe overlaid over a map of the volcanoes of Mexico.

12. Also, Our Lady of Guadalupe as has been determined by Obstetricians, who have examined the image, is pregnant (which is why she is the Patroness of the unborn). Mary's head, by the way, points to Israel, where, of course, she birthed Jesus.

So, what do we have here? We have a cactus fiber tilma that shouldn't still be around. We have an image that is more of a Polaroid or digital image (including images of people in Mary's eyes) that no one can duplicate, much less no one in the 16th century. We have a tilma that should have been blown up. We have an arrangement of stars and of flowers on the dress that no one in the 16th century would have known how to arrange. Not even discussed are the miracles of healing associated with Our Lady of Guadalupe, including one within the first month of the Tilma's existence. But what of the Spanish art images? I believe God designed the image to be similar to Spanish art in order to make sure the Spanish knew who she was. Or, alternatively, God inspired the Spanish artists to paint/sculpture images of Mary as she was to appear in Mexico within 100 years. It does appear that certain areas were painted onto the tilma—areas such as the gold edging on the cape, the fur trim at the wrists, perhaps even the solar rays and the angel.[88] Maybe these were added

---

[88] Philip S. Callahan, *The Tilma under Infra-Red Radiation: An Infrared and Artistic Analysis of the Image of the Virgin Mary in*

by Marcos Cipac de Aquino or some other over-pious artists. However, these areas all now show evidence of flaking. The rest of the actual image of Mary remains perfect, not faded, not painted, unexplainable.

The Real Miracle of Our Lady of Guadalupe

But there is more to this story. As I stated at the beginning of this account, the Spanish were at odds with the Native Americans, as is often the case when a conqueror rules over the conquered. The image of Our Lady of Guadalupe was an image that the Spanish would recognize, and they would even recognize the choice of name of Our Lady. (She could have called herself Our Lady of Tepeyac.) The fact that Mary was sent by God to a Native American convinced the Spanish that the Native Americans were human after all, and that they needed to be treated as fellow humans. The Aztecs, on the other hand, recognized through many coded messages contained as pictographs within the image,[89] which was intentionally given to a Native American (not discussed because of space limitations), that the Aztec creator was the Spanish God, contained within the womb of Our Lady of

---

the Basilica of Guadalupe. II, ser. 3 (Center for Applied Research in the Apostolate, 1981).

[89] See Books of Interest at the end of this book to learn more details of the coded messages.

Guadalupe. As a result, over nine million Mesoamerican Natives converted to Catholicism within a decade. (As five million Catholics became Protestant in Europe, almost twice that many new Catholics were converted and entered the Catholic Church in the Americas.) Thus, the Spanish and the Native Americans viewed each other with a new perspective and so mixed and intermarried. Compare this to the United States, where the Native American's land was taken, and the Native Americans were sent to reservations. In addition, the Europeans in the United States brought over slaves and considered themselves to be better than both the Native Americans and the slaves, building racism into the country's mores. Because of Our Lady of Guadalupe's intervention, little racism has existed in Central America (although recent studies indicate sadly that this may be changing).[90] All peoples were treated with greater equality. (No system is perfect.) If only that were true of the United States, which as a nation still struggles to overcome years of racism.

---

[90] Daniel Zizumbo-Colunga, "Study reveals racial inequality in Mexico, disproving its 'race-blind' rhetoric" (December 13, 2017), at theconversation.com/study-reveals-racial-inequality-in-mexico-disproving-its-race-blind-rhetoric-87661.

Things to Think About

1. Can you imagine walking 13 miles one way to receive catechetical instruction? What extraordinary efforts have you made in your life to learn more about your faith?

2. What the physical aspects of the tilma of St. Juan Diego that are so remarkable and unexplainable?

   As with the Eucharistic miracle at Lanciano and the Shroud of Turin, there are many physical characteristics of the tilma of Juan Diego that are simply unexplainable. The tilma became a miraculous cloak somehow. Its longevity, its resilience, its history, its inherent codes and images, etc., are quite remarkable. However, from a physical perspective it is the polaroid-like nature of the image that makes it scientifically unexplainable.

3. What spiritual aspects of the tilma of St. Juan Diego are remarkable and unexplainable?

Chapter 12

How About a Little Team Spirit?

Over the course of human history, there have been thousands of occasions in which an apparition of God, or a saint, or a relative, or just a person—often referred to as a ghost has been claimed to have been seen. Now you might say, "Come on, Dr. Keys, you don't believe in ghosts, do you?" Well, let's be open about this. Isn't the recorded history of mankind seeing, perhaps, hundreds of thousands of apparitions, evidence that something is happening. Can all those people be making this up? Certainly, we know that some are made up for one reason or another. Shame on them. Yet, most people are sincere. Perhaps individuals can be fooled, or perhaps not. Certainly, one feels uneasy about just believing some "cock and bull" story (or a fish story, or a shaggy dog story, or a whopper) about seeing a spirit. One immediately questions the story teller's sanity. But, logically, just because science cannot explain how some stories can be true, that doesn't mean they are all false. Looking back at chapter 8, one sees how science is gradually accumulating indications that we are body and soul. If one is open to that, then maybe there might be at least one story about seeing a spirit that might be true. Can thousands of stories be due to temporary or permanent insanity of the seer?

Well, what if there were multiple seers? Could three or four seers be sharing in some kind of mass hysteria? Maybe. What if there were more? What if there were 70,000 to 100,000 people seeing an event which no one had absolutely any predisposition as to what would occur? What if the event was witnessed by contrary newspaper reporters, contrary government officials, contrary skeptics, in addition to those who came to see if a promised miracle would occur? What if there were other physical corroborating evidence? Just what would it take for you to believe? That answer is easy. As with all knowledge you must trust in the source(s), whether that be measurable evidence you can evaluate or evidence from people you trust. Fortunately, in the account I am about to relate, there are plenty of skeptics who were there and tell the story!

Before I discuss the evidential cases of apparitions, I would like to first mention, for those who are concerned that Catholics focus too much attention on Mary, that God decides who is to be sent to whomever He wishes to send one to. Thus, in some cases, God sent men on His behalf, such as sending Moses to free the Israelites from Egypt. In other examples, Scripture tells us, He sends angels in His stead, as in case of Lot or to Mary, the biological mother of Jesus. Sometimes, apparitions of people who no longer walk the earth are sent as in the case of Samuel appearing to Saul (1

Samuel 28:7-20), or Elijah and Moses at the Transfiguration (Matthew 17:1-8). So, if God chooses to send Mary to people that is, of course, His choice. To disregard a miraculous event such as an apparition because the apparition is of Mary, the Mother of God, is simply a case of willful blindness, a case of ignoring the evidence. Keep that mind open. Now, let's look at a case with tens of thousands of witnesses.

The Apparitions at Fatima

Every year, over 5 million people descend on the small town of Fatima in Portugal. Why do they do this? Fatima is no resort capital, nor does it have a large university, nor was a point of historical significance—at least not of secular significance. Fatima is, however, the site of six 20th century apparitions of the Blessed Virgin Mary, the mother of Jesus, along with several prophecies which came to be fulfilled, plus a predicted, unexplainable miraculous event witnessed by over 70,000 individuals which occurred on October 13, 1917, so that all may believe. The apparitions were beheld by three young Portuguese children over a course of 6 months. This was a time when Europe was ablaze with war. Portugal was at war with Germany and Austria, and the United States had just entered the World War I in April of 1917. On the eastern front, the Marxist revolution had begun, and the Bolshevik Russian revolution was about to start. Marxism and Socialism was rampant in parts of Europe. Into this turmoil, Mary, the Mother of Jesus, was sent with a message of hope and a call to repentance.

Here is a short account of the pertinent events. The story begins in 1915 and 1916 when an angel appeared multiple times to three children—Jacinta Marto, age 5, Francisco Marto, age 7, and Lucia dos Santos, age 8 (in 1915). The angel asked them to pray and sacrifice for the conversion of sinners. Then, according to the children, on May 13, 1917, a woman appeared to them floating above a small helm oak tree at the Cova da Iria near Fatima, Portugal. What was it this woman wanted? The woman told them, "*I want you to come here for 6 months, on the 13th day, at the same hour. I will tell you who I am and what I want. And I shall return here again a seventh time.*" Then she said, "*Do you want to*

Figure 12.1 Jacinta Marto, age 5, Francisco Marto, age 7, and Lucia dos Santos, age 8 (in 1915).

offer yourselves to God to endure all the sufferings that He may choose to send you, as an act of reparation for the sins by which He is offended and as a supplication for the conversion of Sinners?" In Catholic terminology, this means to unite their sufferings to that of Christ as an offering to God for the sins of the world and the conversion of sinners. What an invitation! An invitation to suffer!

Now you may think, "These are just little children. They are making up stories. What do they know?" And that would be the point. They are far too young to construct such a story, which seemed to be fanciful at best. Besides, what child would make up a story that they will suffer for someone else? So, at this time, you may be thinking, why am I reading this? But, for now, just be open to the account. Through Mary, God gave the world a miracle to help them believe. But why Mary? Why not Jesus Himself? That is the prerogative of God. I can't answer that, but, in history, there are far more apparitions of Mary than of Jesus.

The story continues. The children were told not to tell anyone, but soon after they got home, Jacinta blurted the story to her mother. Word spread. No one, especially the parents and the Catholic Church believed them, which is quite understandable. On June 13, 1917, the three plus 14 of Lucia's friends and a few adults traveled as much as 20 miles to get to the Cova. The woman again appeared in a cloud above the same helm oak tree. Others saw just a cloud. She asked them to come again next month, to pray a specific prayer as part of the Rosary, *"O my Jesus, forgive us our sins, save us from the fires of hell, lead all souls to heaven, especially those in most need."* She instructed them to learn to read and write and told them that Jacinta and Francisco will soon be taken to heaven (Francisco later died April 4, 1919, and Jacinta died Feb 20, 1920). Lucia was to remain, in order to spread a devotion to the Immaculate Heart (of Mary) which would lead others to Christ. The three met with their parish priest the next day. The priest's first thought was, "It might be the devil!" By this time, the children were getting some un-wanted attention. The children began to suffer due to the notoriety and, essentially, there was name calling going on. By now, you might be saying, "This is just Catholic stuff." The question remains however, "Is it true?" It is, after all, truth that we seek.

On July 13, 1917, the children returned to the Cova, although Lucia went begrudgingly. The road to the Cova was packed with curiosity seekers. The Lady arrived as promised and engulfed the children in light. She began her apparition by asking them two things—to come again in August, to pray the Rosary every day for peace in the world and the end of the war. Lucia asked, *"Will you please tell us who You are and perform a miracle so that everyone will believe that you really appear to us?"* To which the Lady replied, *"In October, I will say who I am, and I will perform a miracle all shall see, so that they believe."*

The Lady then showed the children a vision of Hell and stated, "You have seen Hell, where the souls of poor sinners go. If people will do what I tell you, many souls will be saved." She told them that if people *"do not stop offending God, another and worse war will break out in the reign of Pius XI."* Of interest is that the pope at the time was Pope Benedict XV. Pope Pius XI did not become pope until 5 years later. And, yes, World War II did breakout 22 years later. Finally, the Lady asked for the consecration of Russia if there is to be peace (this was prior to the Bolshevik revolution). The consecration was not done, and communism became the rule of the land in what became known as The Union of Soviet Socialist Republics.

To be favored by God to have a visit from the Mother of God was a great difficulty for the children at this point. The children's families lost all their privacy. The press accused the children of fraud. Even Lucia's mother publicly denounced her. Yet visitors continued to stream to Fatima. The fourth apparition which was to be held on August 13, 1917, did not happen as Artur Santos, the magistrate over the town of Fatima, kidnapped the three children and threatened their lives by boiling them in oil if they did not tell him their secrets. Meanwhile at the Cova while the children were being held captive, thunderclaps were heard, along with a flash of light, and the appearance of a cloud over the helm oak. The cloud disappeared and, immediately following, the faces of all those present were bathed in red, white, and blue light.

The Lady knew it was not the children's fault, and on August 19, 1917, the Lady appeared for the fourth time to Lucia and Francisco. Francisco's brother, who was also there, ran home to get Jacinta. The Lady once again asked that they come back next month on the 13th and asked that they pray for the conversion of sinners "for many souls go to Hell *for not having someone to pray and make sacrifices for them.*" The Lady repeated her promise that she would do a miracle in October.

September 13, 1917, arrived. Twenty-five to thirty thousand people were now present. The Lady told the children, *"Let the people pray the Rosary every day to end the war. In the last month, in October, I shall perform a miracle...Joseph will come with baby Jesus to give peace to the world. Our Lord will also come to bless the people. Besides Our Lady of the Rosary and Our Lady of Sorrows will come. God is pleased with your sacrifices...Use half of the money for the litters. And half to help build a chapel."* (Note—the visitors had been leaving donations.)

For the first time, others saw the apparition. It was estimated that about 2/3's of the people who were present saw the apparition. A Monsignor Quaresma from Leiria, Portugal (16 mi away), was there that day and later stated in 1932 in a letter to a colleague: *"There had not been a cloud in the deep blue sky and I too raised my eyes and scrutinized it in case I should be able to distinguish what the others, more fortunate than I, had already claimed to have seen...*

Figure 12.2 Images of the crowd at Fatima before and during the Apparition October 13, 1917.

With great astonishment I saw, clearly and distinctly, a luminous globe, which moved from the east to the west, gliding slowly and majestically through space." He asked his friend what he saw: "*That was Our Lady!*"

Finally, the day arrived—October 13, 1917, the day of the promised miracle. Surely, others will see the miracle, and would no longer harass the children and their families. The adults were worried—what would happen if the miracle did not happen? Would fear and panic rule the day. Would the children be harmed? The children were again pressured to admit they lied. Newspaper journalists arrived from Lisbon and elsewhere. Many gathered ahead of time at Cova, sleeping in the open air overnight. The day began with a light rain which later turned to a hard rain. It has been estimated that 70,000 to 90,000 were present.

Noon came. Suddenly the clouds dispersed. The sun became as pale as the moon. St. Joseph appeared (waist on up) to the left of the sun holding the child Jesus. Our Lady appeared in blue and white as Our Lady of the Rosary, then later appeared in the clothing of Our Lady of Sorrows, and finally as Our Lady of Mount Carmel. All were bathed in various colors of light.

Then suddenly, the Miracle of the Sun began. The Sun started rotating, then began dancing in the sky, before finally plunging toward the earth. People were on their knees and were shouting. Many thought this was the end of the world. After about 10 minutes, the sun resumed its place in the sky. The people there, whose clothes were soaked by the earlier rain, and whose knees were muddied by kneeling on the ground during the amazing events, suddenly found their clothes dry and clean, the mud stains being gone.

Evaluation

The first impression is that this story sounds so fanciful, that it can't be true. But let's look a little closer. Let's imagine that the children made up this story. How could three children, ages 7, 8, and 10 years old, orchestrate such an event? It's one thing to say that a Lady appeared to them standing on a cloud above a helm oak tree. It's another to say that the Lady wants them to come back for on the 13th of the month for five more months. Furthermore, the children stated that there would be a miracle on the 13th of October. That's putting a lot of pressure on them to perform! These same children would have to have predicted that there would most likely be a Second World War. They would need to know that the next Pope would choose the name Pius. They would need to be able to see the danger to the world of Russia's becoming communist. Certainly, they had no idea of the state of things in Russia, much less what communism really means. Why would these children want to undergo the sacrifices and sufferings that they endured? Was this a bad idea, that went downhill? Was this a potential "Breaking Bad" scenario, spiraling ever deeper into the muck of lies? Well, maybe their parents put them up to it? Get real! Even the parents didn't have the ability to accomplish this great deception. Besides, with the exception of Ti Marto, the father of Francisco and Jacinta who came to believe the children by August 13, 1917, the parents didn't believe or support the children. The Catholic Church at first thought it was the work of the devil, then gradually came to a let's look and see attitude. The Catholic Church did not promote it, perhaps, because the collapse of such a deception would certainly be detrimental.

Well, maybe it just didn't happen. Perhaps their account of the happenings in Fatima in 1917 is not an account but is just a story. The pictures above of those at the Cova are just pictures of people at some air show, watching the air show in awe and wonder. Well, there is more evidence than just word of mouth and a few undocumented pictures. Portugal was not a clerically friendly place. After the deposition of the King Manuel II in 1910, the new republic imposed severe anticlerical measures.[91] Both the government and the newspapers deplored the Church. That is why Artur Santos, the magistrate of Fatima, for political reasons, so wanted the apparitions in his town to go away and become a forgotten issue. So, because the mood of those running the country was so atheistic in nature, the government officials and the newspapers were there that day in October to refute anything that might be said about a supposed miracle. Surely, the secular

---

[91] Wikipedia, "5 October 1910 revolution," at en.wikipedia.org/wiki/5_October_1910_revolution#Anticlericalism.

newspapers newspapers—*O Dia, O Seculo* (Lisbon) and *Ordem* (Oporto) would record the event and expose the deception of the three children.

But that was not to be the case. The miracle was not just seen at the Cova, but eyewitnesses 20 miles away reported seeing the sun dance in the sky. The sun's dancing in the sky was not reported in Lisbon, Paris, London, etc. But that would be a misunderstanding of how these things work. This is a miracle after all. One shouldn't think that the physical sun actually danced in space, wreaking havoc on the orbits of the planets, etc. Similarly, one need not consider the events of the star of Bethlehem leading the astrologers to Mary, Joseph, and Jesus to be an actual astronomical event. Nor must one have to consider that Joshua's asking God to have the sun stand still at the siege of Gibeon (Joshua 10:13) actually kept the sun from "moving" (or alternatively, the earth from rotating). Miracles alter the laws of physics and associated realities only at the site where the miracle occurred. Yes, the people within 20 miles of Fatima (nearly) all saw the sun dancing in the sky. (Note: this phenomenon of the sun has been reported a number of times in the last 40 years at Medjugorje in Herzegovina. You can google the videos taken and judge for yourself.) And, yes, they certainly suddenly had dry, non-muddy clothes, which is miracle enough.

So, what have others said about this event? I have chosen two comments to compare. One is from the noted atheist,

Richard Dawkins, PhD, the other from José Alves Carreia da Silva, Bishop of Leiria from 1920-1957.

Richard Dawkins: "*It may seem improbable that 70,000 people could simultaneously be deluded, or could simultaneously collude in a mass lie...but any of those apparent improbabilities is far more probable than the alternative: that the earth was suddenly yanked sideways...with nobody outside Fatima noticing it.*"

Here Dawkins admits that 70,000 people saw some event, but then throws away that testimony and dismisses Fatima because it does fit into his worldview. He also states that the "earth was suddenly yanked sideways" which was not what anybody said. The people said that it appeared the sun was dancing in the sky. That is not the same thing. His reality is strictly that of a scientific reality, thus causing him to dismiss the evidence of 70,000 people. Thus, he never ponders what this could mean. 70,000 people cannot be deluded, he says, therefore, he admits there is a truth in their common witness. Yet, he knows the physics of the sun dancing would have ramifications. So, with "willful blindness" he dismisses one truth and accepts the other, without realizing there might be another understanding—that of a truly miraculous event which somehow remains local. But Dawkins doesn't believe in miracles (at least not till we get to chapter 13) and will never allow himself to understand if indeed a miracle is the answer. One, also, has to remember what Dawkins said about the Argument of Personal Incredulity, which he used against some Christians back in chapter 6. Here Dawkins admits there is evidence, but he doesn't believe it because he is incredulous!

In contrast, Bishop José Alves Carreia da Silva: "*The children long before set the day and the hour at which it (the miracle) was to take place...* ***This miracle was not natural.*** *It was seen by people of all classes, members of the Church and non-Catholics. It was seen by reporters of the principal newspapers and by people many miles away*".

Here in his response, he simply acknowledges what happened. No one can scientifically explain how such a phenomenon could occur, much less at a predicted date and time. He accepts the accounts of eyewitnesses, including that of antagonistic newspapers. His account is actually more scientific than that of Dawkins and so many other scientists. There is evidence that is anomalous. He doesn't just throw it away as Richard Dawkins would have one do. Again, 70,000 people cannot be deluded, and the Miracle of the Sun was but one of the miracles observed. Maybe someday, someone will come up will an irrefutable, evidential answer as to how these predictions and events can come about. This should not destroy one's faith, for private revelations are not required to be believed, but admittedly, can be faith builders. But for now, the evidence is in the other direction—multiple miraculous events did happen as predicted. And, in the meantime, I would suggest that we all should pray, repent, and change our ways, for the world just isn't the way it should be. It all starts with each individual.

The Incorruptible Bodies

Well, here I go again with more Catholic miracles. The Catholic Church has had many miracles over its 2000-year history, and so many more are available today to see. Once again, science and medicine have no explanation of the miracles I am about to describe. Incorruptible bodies are bodies which fail to decay as one would expect after death. The bodies of the individuals who have remained incorrupt have a recorded personal history of being extremely devout Catholics or Orthodox Christians. Unlike miraculous healings, incorruptible bodies seem to be a Catholic or an Orthodox event (Both Churches historically can trace their origins to Jesus and the Apostles). No other religious or atheist group speaks of their members as having incorrupted bodies (unless the body was known to be stored in a dry, arid location). Once again, they can be considered doctrinal, because they remind us of the doctrine of the Resurrection of the Dead. As you will see, we hope our glorified bodies will be in better shape than these, but one cannot deny that these bodies shouldn't be here at all.

For many the hang-up with the Incorruptibles lies in the fact they often tend to show signs of corruption. So, let's look at the definitions of Incorruptibles and compare it to other preserved bodies which occur in the world. There are three categories of preserved bodies:

1) **Accidently preserved** — Some natural environments are ideal for the preservation of human flesh and bones. Examples of this are being buried in dry, hot sand, or in lava, or in areas of high radioactivity. A major requirement is that moisture doesn't reach the body. These bodies are discolored, wrinkled, distorted, skeletal looking and rigid with limited or no elasticity.[92] Once bandages are removed, they soon become odiferous and begin to decay.
2) **Deliberately Preserved** — As it sounds, these bodies were intentionally preserved by embalming. The ancient Egyptians used to remove the internal organs and place resin or other substances inside in order to minimize decay. These bodies tend to look very similar to the natural preserved bodies.
3) **Incorruptibles** — Bodies which are incorruptible can be dated back to the early centuries after Christ. These bodies should have decayed years ago. Often the environments in which the bodies reside are moist, which would be extremely deleterious to preservation. Incorruptible bodies are often found near

---

[92] The Incorruptible Saints, at www.roman-catholic-saints.com/incorruptible-saints.html.

other corrupted bodies. There is no scientific explanation why these bodies do not decay while adjacent bodies do. Some of the characteristics include:

- Incorruptibles are typically found lifelike, moist, flexible, and contain a sweet scent that many say smells like roses or other flowers, for years after death.
- Incorruptibles are almost never embalmed or treated in any way due to the beliefs of the religious order from which the person came.
- Incorruptibles remain free of decay, some for centuries, despite circumstances which normally cause decay such as being exposed to air, moisture, other decaying bodies, or other variables such as quicklime, which is typically applied to a corpse to accelerate decomposition.
- Incorruptibles many times contain clear, flowing oils, perspiration, and flowing blood for years after death, where accidental or deliberately preserved bodies have never been recorded to have such characteristics.
- Other partial incorruptibles have been found throughout the centuries where certain parts of the body decay normally while other parts such

as the heart or tongue remain perfectly free of decomposition.

Many people express concern when they find that many incorruptibles either were never totally incorrupt, or that tissue that was incorrupt has begun to become corrupt. In fact, in order to make a more pleasing presentation, many of the incorruptibles have had wax placed over what had become corrupted. For instance, St, Bernadette Soubirous (d. 1879) has a thin wax coating on her face, making her appear to be a beautiful 24-year-old woman, who is simply sleeping. Does this mean that she is not incorrupt, that the caretakers of her body are trying to fool people? No, if you look at the definitions of the incorruptibles, you will see that he incorruptibles are found a certain way. Over a period of time, due to the effects of being exposed to moist air, smoke from candles, various wear and tear, the bodies may begin to show damaging effects. But even a face whose skin has turned dark still has skin and muscles which shouldn't be there. Normally, so called black putrefaction occurs in 10 to 20 days.[93] All soft tissue if exposed to the air decays in three to seven weeks, not decades! Wooden caskets will delay the process, maybe even as long as 50 years. Bones with all the calcium

[93] "Stages of decomposition" (June 25, 2020), at australianmuseum.net.au/movie/stages-of-decomposition.

they contain can remain, depending on conditions, 300 years but will eventually disappear.

So, when an incorruptible body is found even possessing a tongue, such as St. Anthony of Padua (a noted orator), or the hands and arms, such as St Francis Xavier (who baptized tens of thousands of people), it is still an unexplainable occurrence by science. However, most incorruptibles had much of their bodies intact when discovered.

Why are they discovered anyway? The Catholic Church considers bodies of the saints to be relics. Relics would certainly be another topic worthy of discussion, but, for now, let me say that there are many well-documented examples of healings that have occurred when a person needing healing comes into contact with the relic of a saint. Now, you may not believe the stories of healings, and I understand that. Just another thought to remain in the back of your mind, why would someone make that up, and then get others to back their story? Unfortunately, there are charlatans out there. There is Biblical evidence. One Biblical example is that of a man who came to life when his body touched the bones of Elisha. Regardless of the bad guys in the world, there are many modern examples of healings that are well documented.

For now, let's set aside the question of relics, let's look at a fairly modern case of an incorrupt body that I have already mentioned. As noted, as is generally the case, it was discovered when the Church sought to obtain token body parts of one of the 19th century's greatest saints—St. Bernadette Soubirous.

St. Bernadette Soubirous

Bernadette Soubirous was born January 7, 1844, near Lourdes, France. At the age of 14, Bernadette began to see apparitions of the Blessed Virgin Mary, who identified herself as the Immaculate Conception at a cave near a garbage dump at Lourdes, France. The Blessed Mother asked that a chapel be built there. Soon, waters from a spring arose there. After overcoming strong skepticism of the Catholic Church, a chapel was built, and well-documented healings

Figure 12.3 Bernadette Soubirous in 1858

occurred for those who bathed in the 54° F spring waters. The healings remain a continuing phenomenon at Lourdes.

Bernadette had always been a somewhat sickly child, suffering from cholera as a toddler and later from asthma and digestive troubles as a child. As an adult, she entered a convent in Nevers, France, became a postulant, but subsequently protracted tuberculosis and died at the age of 35 in 1879. Her body was placed in a double coffin of lead and oak, which was sealed in the presence of witnesses and soon placed in a crypt in a small chapel dedicated to Saint Joseph, which was within the confines of the convent.

By 1909, due to Bernadette's reputation for saintliness and due to miracles attributed to her intercession, an episcopal commission sought to examine her body and to possibly obtain relics. This is, of course, a big event, and on September 22, 1909, with a number of people present, including the Monsignor Gauthey, the Bishop of Nevers, the church tribunal, the mayor of the town, 3 official witnesses, 2 doctors (Jourdan and David), 2 stonemasons, and 2 carpenters, the coffin was opened for the first time since 1879. There was no tint of odor. The body was perfectly preserved. In the words of the doctors:

> *"The coffin was opened in the presence of the Bishop of Nevers, the mayor of the town, his principal deputy, several canons and ourselves. We noticed no smell. The body was clothed in the habit of Bernadette's order. The*

habit was damp. Only the face, hands and forearms were uncovered. The head was tilted to the left. The face was dull white. The skin clung to the muscles and the muscles adhered to the bones. The sockets of the eyes were covered by the eyelids. The brows were flat on the skin and stuck to the arches above the eyes. The lashes of the right eyelid were stuck to the skin. The nose was dilated and shrunken. The mouth was open slightly and it could be seen that the teeth were still in place.

The hands, which were crossed on her breast, were perfectly preserved, as were the nails. The hands still held a rusting rosary. The veins on the forearms stood out. Like the hands, the feet were wizened and the toenails were still intact (one of them was torn off when the corpse was washed). When the habits had been removed and the veil lifted from the head, the whole of the shriveled body could be seen, rigid and taut in every limb.

It was found that the hair, which had been cut short, was stuck to the head and still attached to the skull -- that the ears were in a state of perfect preservation -- that the left side of the body was slightly higher than the right from the hip up. The stomach had caved in and was taut like the rest of the body. It sounded like cardboard when struck. The left knee was not as large as the right. The ribs protruded as did the muscles in the limbs. So rigid was the body that it could be rolled over and back for washing. The lower parts of the body had turned slightly black. This seems to have been the result of the carbon of which quite large quantities were found in the coffin."

In witness of which we have duly drawn up this present statement in which all is truthfully recorded.

Nevers, September 22, 1909

Drs. Ch. David, A. Jourdan.

As you can read, the body was not, as sometimes reported, as if Bernadette had just gone to sleep the day before. But you have to realize that what should have been left would only have been bones. The body was washed, and some salts placed on the skin, and placed her in a new casket. Having been exposed to the air, the skin began to darken.

This is not the end of the story. There's more. The body was again exhumed in 1925. This time the Church brought in two atheist physicians from Paris, Drs. Talon and Comte, to do the examination. Dr. Comte wrote *"The body is practically mummified, covered with patches of mildew and quite a notable layer of salts, which appear to be calcium salts. ... The skin has disappeared in some places, but it is still present on most parts of the body."*[94]

---

[94] Wikipedia, "Bernadette Soubirous," (March 19, 2018), at en.wikipedia.org/wiki/Bernadette_Soubirous

This was followed by a third exhumation on April 3, 1928. This time Dr. Comte was asked to obtain relics. He reported,

> *"What struck me during this examination, of course, was the state of perfect preservation of the skeleton, the fibrous tissues of the muscles (still supple and firm), of the ligaments, and of the skin, and above all the totally unexpected state of the liver after 46 years. One would have thought that this organ, which is basically soft and inclined to crumble, would have decomposed very rapidly or would have hardened to a chalky consistency. Yet, when it was cut it was soft and almost normal in consistency. I pointed this out to those present, remarking that this did not seem to be a natural phenomenon."* [95]

None of this would have been possible had the body been mummified. Further evidence of the reality of this miracle is that Dr. Comte gave up atheism and became Catholic.

But you may say, "What's the big deal? There are a few other bodies in the world that also have not corrupted." Yes, that is true, however, those bodies have been found in very unique dry arid conditions. The fact that other bodies in unique situations have not decayed should not interfere with

---

[95] *Bulletin de L'Association medicale de Notre-Dame de Lourdes, Issue 2, (1928).*

the recognition of miraculous events for ordinary bodies, in ordinary situations.

Other noted happenings

This chapter covered just one example of a well-documented apparition, that of Fatima. There are many more, one could discuss. There have been 2600 approved apparitions since 40 CE. In the 20th century, there were 7 apparitions approved. In one apparition in Zeitoun/Cairo, Egypt, beginning in the year 1968, over 2 million persons saw apparitions of Mary. Even the Egyptian President Nassar saw the apparition. Those apparitions were recorded and displayed by the Cairo TV stations. There are also, of course, many other examples of well-documented incorruptible bodies. Other physical phenomena exist—dried blood of certain saints sitting on the shelf suddenly liquefies year after year on certain dates, certain people have experienced inedia—the ability to live solely on the Eucharistic host and Most Precious Blood. One of the most recent was Marthe Robin of Châteauneuf-de-Galaure in Southeastern France. She became bedridden at the age of 21 in 1923 and lived the rest of her life soley on the Eucharist without losing weight! She died in 1981. She had been examined many, many times by physicians who documented her inedia. And 1981 is just recent history. Of course, she is not the only person to have lived that way. Strange things do happen. Many other strange, well-documented phenomena could also be discussed—stigmata, bilocation, levitation, bleeding, or olive oil producing statues (even ones evaluated by computerized tomography), statues that mysteriously move their arms or eyes.

In the next chapter we will summarize our journey and look at this last example in more detail through the eyes of, who else, but Richard Dawkins. So why wait? Onto chapter 13!

Things to Think About

1. Can you imagine the anguish the three young children of Fatima must have endured? They must have been under tremendous pressure. As a young child, could you have withstood onslaught of negative attention, they must have felt?

2. *The Miracle of the Sun* was seen within a 20-to-25-mile radius of the Cova de Iria near Fatima. The rest of the world did not see such an event. Is it possible that such a thing physically happened to the Sun?

3. Many of the bodies of saints have decayed somewhat since they have been discovered and put on display. Some have been restored through the use of wax and other means. Does this mean they really aren't

incorruptible? Does this mean there has been no paranormal event going on?

Chapter 13

Removing the Blinders

I began this book by pointing out what I hope is obvious to all by now—to understand anything, one must seek the FULLNESS OF REALITY. The problem is, of course, that no one will ever understand all of Reality. We are limited in our capacity to understand, in the amount of knowledge we can ascertain, and in our own biases which keep us from seeking all truth. The best we can do is to make use of knowledge from sources, which are, to the best we can determine, trustworthy. We need to be open to truth, and to seek congruence when two different truths apparently contradict each other.

The physical sciences are a wonderful thing. They examine the What and the How of existence. As such, the physical sciences are lacking because, by their nature, they can only look at part of Reality. Religion and Philosophy on the other hand try to ascertain the Why of existence. So why would it be so surprising that these different groups sometimes reach different conclusions, particularly when they try to answer questions from each other's realm using solely their own knowledge.

I remember when I saw my first horse race. I asked my father, "Why do the horses wear blinders?" He told me the blinders permitted the horses to only see straight ahead, and, as a result, they would not be distracted by the other horses. It may seem contrary, but this is actually a good idea for our scientists and our theologians. Scientists look for the WHAT and the HOW from the materialist world. That is their field of study. Theologians look for the WHY and the WHO from the non-material, transcended world. That is their field of study.

On the theologian side, when science shows evidence which directly contradicts what has been traditionally taught by theologians, theologians must remember they are trying to understand revelation—the WHO and the WHY, not science—the WHAT and the HOW. They must also consider the truth of science while remembering that science often changes its mind.

On the scientist side, it is good to try to explain physical phenomena by using known scientific laws, but when science fails to explain phenomena, scientists should look at all sources of answers. Scientists in good faith should realize they have been intentionally wearing blinders with the intent of understanding the physical side of existence through only the physical. They must correspondingly remember that they are simply looking for the WHAT and the HOW, and not be declaring there is no WHO and WHY to be concerned about! They must not accept what wild assertions such as probabilities like 1 in godzillion only so they can keep a higher power, namely God, out of the picture. If the evidence points toward a non-physical side of Reality, they should incorporate that information into their understanding of the FULLNESS OF REALITY. Albert Einstein said it best, "A man should look for what is, and not for what he thinks should be." Scientists, fundamentalists, and all seeking Truth should follow that creed. But so many do not.

We see the blinders on in writings such as Stephen Hawking's book, *The Grand Design*. There we find Hawking from the beginning stating there are no miracles and philosophy is dead.[96] He could, perhaps, have said, we are going to study the "Grand Design" of Nature only from a physical point of view. But to say there are no miracles simply must mean that Hawking never investigated the miracles found in the world—that would be a form of willful blindness. Furthermore, to say Philosophy is dead does not really mean that Philosophy is meaningless and is of no use; rather, it means that Hawking didn't agree with or understand Philosophy (not to say all Philosophy is correct). After all, for Hawking, only particles, waves, forces, etc., exemplified what makes up Reality.

---

[96] Stephen Hawking, and Leonard Mlodinow, *The Grand Design* (New York: Bantam Books, 2010), 38.

Richard Dawkins, on the other hand, states that miracles are real. Miracles for Dawkins are just extreme examples of probability theory. In chapter 12, I noted that Richard Dawkins has discussed the concept of a statue of the Virgin Mary waving its hand. Here is what he stated in greater detail:[97]

> *"A miracle is something that happens, but which is exceedingly surprising. If a marble statue of the Virgin Mary suddenly waved its hand at us, we should treat it as a miracle, because all our experience and knowledge tells us that marble doesn't behave like that...*
>
> *"In the case of the marble statue, molecules in solid marble are continually jostling against one another in random directions. The jostlings of the different molecules cancel one another out, so that the whole hand of the statue stays still. But if, by sheer coincidence, all the molecules just happened to move in the same direction at the same moment, the hand would move. If they then all reversed direction at the same moment the hand would move back. In this way it is possible for a marble statue to wave at us. It could happen. The odds against such a coincidence are unimaginably great but they are not incalculably great. A physicist colleague has kindly calculated them for me. The number is so large that the entire*

[97] Dawkins, *The Blind Watchmaker*, 159.

age of the universe so far is too short a time to write out all the naughts! It is theoretically possible for a cow to jump over the moon with something like the same improbability. The conclusion to this part of the argument is that we can calculate our way into regions of miraculous improbability far greater than we can imagine as plausible."

So, Richard Dawkins believes in miracles, but not in miracles originating from God. Instead, miracles originate out of highly unlikely probabilities. For him, if you can calculate the probability, it is plausible! It might happen! It is utterly incredible that he would believe this. Math is abstract. It is not physical. Math is heavily used in the theoretical models of the physics world. However, there is always a point where one comes to find that the math only approximates reality. For instance, nothing physical is truly circular. Objects at best only approximate circles. Because someone can estimate that there is a probability of 1 in a trillion, trillion, trillion, trillion, trillion, trillion, trillion, trillion, trillion, ad infinitum, doesn't mean that whatever was calculated is ever going to happen.

To help you see what I mean, let me give you an example. In the 5th century BCE, there was a clever Greek philosopher named Zeno of Elea, who liked to pose paradoxes in which an everyday activity seemingly became impossible to happen. Here I will give a very Zeno-like paradox which is solved once one understands the reality of physics versus the abstractness of math. A runner is in a 100-meter race. The gun sounds and the runner is off and running. Soon he is 50% of the way to the finish line (50 meters). Next, he will cover 50% of the remaining distance and will be 25 meters away from the finish line. Then, he will cover 50% of the remaining distance and will be 12.5 meters from the finish line. Continuing, he will cover 50% of the remaining distance and will be 6.25 meters away from the finish line. Persevering, he will cover 50% of the remaining distance and will be 3.125 meters away from the finish line. Then, he will cover 50% of the remaining distance and will be 1.625 meters away from the finish line. Mathematically, by looking at the runner's progress as 50% of the remaining distance, the runner never reaches 0.0 meters from the finish line. In this analysis, the runner cannot finish the race.

Well, obviously, there is something wrong with the analysis. However, conceptually, it would seem there is nothing wrong with the analysis. Every statement would seem to be true. That is why it is an apparent paradox. But here is where the abstractness of the math falls apart, and the reality of physics sets in. One cannot keep going to smaller and smaller increments. Quantum theory states that there is a finite minimum distance one can move. That distance is called the Planck length and is equal to 1.6×10^{-35} meters. (Just like I couldn't put into a mental image a fathomably large distance when speaking of the universe, or the probabilities of fine-tuning, I cannot put into a mental image a fathomably small distance the concept of the Planck length.) Certainly, at that point, a runner cannot just move to 0.8 x 10^{-35} meters from the finish line. He must cross the finish line, or he must simply stop. I vote for crossing the finish line.

Dawkins explanation of miracles, then, as being an unbelievable statistical UNLIKELIHOOD which happens according to the world of mathematics is simply not realistic. As you intuitively know, a cow is never going to jump over the moon out of some statistical variation. A statue is never going to move its hand out of some statistical variation. If a statue were to move its hand, it is not due to statistical variations. It is due either to shenanigans, or due to a Divine purpose.

So, I can be confident and boldly state, "A cow is never going to jump over the moon" even if some misguided probability states it could happen. That would be NEVER! A statue is never going to move its hand and wave out of some statistical vibration. That would be NEVER! Statistically, all of a sudden all the air molecules could move to the other side of the room where you are reading this book and you would suddenly die from suffocation. Don't worry, that is never going to happen. That would be NEVER! This is a complete misunderstanding of the physical world and the understanding of probabilities in the Real World.

Here is another example, but with a resultant different point of view. The example is the Eucharistic Miracle of Lanciano which we saw in chapter 10. I can easily say that a consecrated Host is never by statistical variation going to transform itself from bread into true heart muscle with nerves and blood vessels, which also, somehow, remains attached to the bread of which it was originally composed. Furthermore, that muscle-bread combination could never still be around according to the laws of physics and the reality of biology after 1300 years, particularly since it has been untreated with preservatives and has been exposed to air! The only rational explanation for this occurrence, if you open your mind to all of reality, is that a force beyond nature has acted upon it and remains acting on it. Remember, this unexplainable, physical (miraculous) event remains ongoing today, and is not the only such Eucharistic miracle in the world. All this is way beyond Dawkins' physical statistics.

Let's leave the world of religious miracles alone for now and review some of the incredible physical phenomena we have already discussed. Remember in the Big Bang the odds of the Universe's entropy being so low was 1 in a godzillion? Remember all those constants associated with the strength of the various forces, the ratio of the mass of electrons to protons, etc., having to be so exact or we wouldn't exist. Remember how an inflationary period with no physics basis had to occur for a precise period in order to have a viable universe. Remember that the same Universe must have had a cosmic evolution perfect for our Earth, to be anthropic. Remember all the information that must contained in the very first living cell, in order that it could replicate itself, create new proteins, have metabolic control, etc. Remember that in evolving from a cell to our present world full of life, suddenly there would be unexplainable developments in a cosmic fraction of a second. Statistics would say that couldn't happen, but it did.

Then remember how Dawkins gave us the clue as to how this all could happen when he proposed a computer could reproduce the phrase *Methinks it is like a weasel*. He showed that the whole process could indeed be sped up if the process was somehow observed by an outside entity who knew what the final outcome should be and selected the approproiate mutations along the way which would achieve the desired result.

So what does theology say about this? It is no surprise that I am Catholic and can turn to the Church for guidance in such matters. The guidance we see in Dawkins' *Methinks it I like a weasel* is called Divine Providence. The Catechism of the Catholic Church (a great book to read whether you are Catholic or not) states in paragraphs 301 and 302:

- **CCC 301** With creation, God does not abandon his creatures to themselves. He not only gives them being and existence, but also, **and at every moment**, upholds and sustains them in being, enables them to act and brings them to their final end.
- **CCC 302** Creation has its own goodness and proper perfection, but it **did not spring forth complete**

> from the hands of the Creator. **The Universe was created "in a state of journeying"** (*in statu viae*) toward an ultimate perfection yet to be attained, to which God has destined it. **We call Divine Providence the dispositions by which God guides His creation toward this perfection.**

Even Quantum Physics and the Catholic Church agree along this thought:

- **Quantum Physics** — Nothing **exists** without being observed.
- **CCC 320** God created the universe and **keeps it in existence** by his Word, the Son "upholding the universe by his word of power" (*Heb* 1:3), and by his Creator Spirit, the giver of life.

God is the big observer. He knew me when I was in my mother's womb.

Do atheistic scientists see a great intellect behind all of science and of life? Of course, they do, but they refuse to call it God. I mentioned in Chapter 2 that Theologians, while focusing on the WHO and the WHY, do speak in an unscientific manner, such as in regard to the intentionality and order of Creation, on the HOW of creation. Science has, in its exploration of the HOW of reality, given us a glimpse into the WHO of REALITY. Scientists have presented all kinds of probabilities which, if it were not that the alternative was the WHO—that is GOD—they would have rejected long ago. For instance, no sane physicist would ever accept that since there is a small, but finite probability, that all the air molecules in a room could go to the opposite side of the room for a while, leaving one gasping for air, that it will actually happen.

Atheistic scientists do see, but refuse to admit, that there is a providence that allows all the unlikely things we have been discussing to happen, overcoming the most horrendous of probabilities. Here are a few examples. I have already mentioned that Steven Hawking wrote a book about the Universe, about quantum theory, about relativity. Note, he called it "The Grand Design." Why call it a design? The answer, because the Universe looks like it is designed—fine-tuned and all. For Hawking, if the Universe has a law of gravity, then it will create a Universe from nothing. He believes that M-theory and the practically infinite existence of different histories of the Universe will explain everything. Of course, the M-theory is not there yet, and it would have difficulty explaining the Eucharistic miracle at Lanciano and other mysteries in the world.

Francis Crick, a Nobel Prize winner for his work on DNA,[98] has stated: "*Biologists must constantly keep in mind that what they see was not designed, but evolved.*" What is he

---

[98] Crick, Frances (8 June 1916 – 28 July 2004)

afraid of? He sees obviously sees a design. Anyone studying molecular biology should find it impossible not to see design. I would wonder if one could pass an exam in a molecular biology class without knowing the design/ manufacturing process for proteins in a cell. The only question is who designed it—Mother Nature (who doesn't exist) in her random acts and cumulative approach, or God in His Divine Providence. At least, God's existence is supported by His miracles.

And let's not forget our overzealous evolutionary biologist, Richard Dawkins. He has stated, "Biology is the study of complicated things that give the appearance of having been designed for a purpose."[99] It is just so hard to miss the intent behind a cell's development.

Of course, one could go on and on into every field of science, whether it be biology, medicine, forestry, ecology, meteorology, geophysics, etc. The world is utterly complex, yet when man doesn't get in the way, it all seems to fit like a glove. So, if every one sees it, why are they afraid to admit the awesome design and intricacies, even if they don't know for sure who the designer is?

Now, perhaps it appears that I am only addressing the willful blindness of some scientists and some theologians. That is far from the truth. Scientists and theologians represent but an extremely small minority of the population.

---

[99] Dawkins, *The Blind Watchmaker,* 1.

However, it is from them that we get much of our information about the world and heaven. Remember, at the beginning of this book, I mentioned that all knowledge comes from sources that we trust—those sources being primarily our senses, our parents, friends, family, teachers, scientists, ministers, etc. How much we believe anything depends on how much we trust the source. All too often, realizing how utterly complex life is, we either ignore completely or trust without any skeptism the information being presented to us. We must both ask the HOW and the WHY questions. Of these, the WHY is the more important, because the WHY gives meaning to life.

In the end, Science and Faith-based views of reality are very similar in that they both look to evidence to determine Truth. Science looks to experiments to obtain evidence of Truth. However, their evidence is never exact, but always has some error in measurement. Furthermore, as I have shown, intrinsic to their measurements is the reality that all interactions are subject to probabilities due to quantum effects. Thus Science Truth cannot be known absolutely. Add in non-physical sciences such as Psychology and social sciences which deal with human behavior and, again, one can only obtain probabilities of a person's behavior. The way a person behaves will depend upon their genetics, their environment, and their ability to make good or bad choices due to their free will. Faith-based views likewise seek evidence. They look to revelation for evidence. Sometimes, they look to writings which they believed are inspired by God, other times they look to events which can change a person's heart, or to miraculous events. From this, they can come to know God, but they cannot ever fully know God.

All of reality, whether from Science or Faith, has fuzzy edges. Reality is a mystery, of which we can never know the complete Truth. The big mistake we make is to ignore one set of Truth because it appears to contradict another set. The narrow Partial Truths of Science and of Faith obscure the fact that they are both looking at different facets of the complete Truth. Always, always, Truth cannot contradict Truth. The two realities must come to a point of union.

So let's review where we have been:

- From nothingness suddenly arose our universe. If you choose to believe that the Universe was burped out of some universe generator, there was nothing where our Universe lies—no dimensions, no energy, no particles, no time, simply naught. And isn't the burping out of the parent universe generator sound just similar to the ancient Babylonian myth where the Universe comes from the gut of the goddess Omoroca who was slain by the god Belus? (See chapter 2.) There is no evidence for either the Babylonian myth or for the Multiverse universe generator.

Alternatively, one could believe that a personal God intended to create the Universe out of nothing for the sake of man, so we can share in God's life. We find evidence for this in the revelation of God in Scripture, in the miracles we encounter, in the body and soul nature of us as human persons, in love and free will of mankind.

- Against all odds, we find that the Universe and its forces are fine-tuned, not only to exist but to be anthropic for man's survival. We can believe we hit the 1 in a godzillion lottery. Such numbers are so large, it is impossible to fathom their magnitude. Let me give you a visual which may help you understand. Let's take the sport of baseball, but it could be hockey, soccer, football, or non-sports activities like driving a car across the country. In all cases, we will utilize the same blindfolded man. Now for the comparison. I maintain, it is far more likely for a blindfolded man situated at the plate in a baseball game, to swing a bat, not even knowing when the pitcher is about to pitch, and hit a home run than for the Universe to be so fine-tuned. Repeat this for all the different fine-tunings we saw in chapter 3.

Alternatively, instead of believing the baseball analogy could actually happen, one could believe that God created the Universe and set its laws, as a lawmaker should, to achieve His goals as intended.

- Against all odds, we find our planet to be the only planet science has found to be in all the required habitable zones with the necessary supplies of minerals for a human civilization, now spinning at an ideal rate of 24 hours a day, with a sun putting out just the right heat necessary for man's existence, and on and on. Once again, the blindfolded man, not knowing when the pitch was to be thrown, has hit multiple home runs, to achieve such an unlikely feat.

Alternatively, one can believe that a personal God under Divine Providence created the Earth for the sake of man so we can share in God's life.

- Against all odds, we find that life was created where there was no previous life, that all the necessary molecules for life came to together at the right moment, with the information necessary to recreate itself, make and repair all the proteins necessary, metabolize energy, and become encapsulated by an outer membrane in a time frame which statistics would say is far, far too unlikely to happen in a god-zillion universe lifetimes. Once again, the blindfolded man, not knowing when the pitch was to be thrown, has hit another home run.

Alternatively, one can believe that a personal, all powerful God under Divine Providence used His creation to create life on Earth for the sake of man so we can eventually share in God's life.

- Against all odds, we find that life developed in utterly complex ways, often in extremely short periods, that all life systems—ecology, photosynthesis, molecular biology, our bodies, etc., have incredible design as part of their essence (beyond the capabilities of man to design) which developed through random actions which were adjusted by cumulative selection over time. During the Cambrian Explosion, the rate of evolution dramatically sped up. Not only did one specie develop eyes, but nearly all did, along other developments of exoskeletons, endoskeletons, etc. To simulate that, we now have multiple blindfolded men, using the experience of the one blindfolded man, swinging away, hitting home run after home run.

Alternatively, one can believe that a personal, all-powerful God under Divine Providence used His creation and His methods to develop life on Earth for the sake of man so we can eventually share in God's life.

- Against all odds, we find that *Homo sapiens* suddenly decide to make drawings, bury the dead, have empathy, love, greed, lust, free will to do good or evil, and to act in a way beyond that of instictive nature. Here, the blindfolded man suddenly decides to go against all his life experiences and swing in his own unique way, yet still hits home run after homerun. In addition, you can join Richard Dawkins and others who believe computers will one day be gathering around the table as persons discussing their origins as simple microprocessors. Do you really believe that can happen?

Alternatively, you can see that mortal man while in the body of a *Homo sapien* is made into the image and likeness of God when God infuses an immortal soul into the being of a man and a woman. This allows man and woman to be able to choose to love God, something a being with no free will could ever do, and allows one through faith and reason to come to know God. From these two individuals all future human persons descend.

- Against all odds, certain statistical variations (or miracles) happen periodically which defy all known laws of science—some come and go, and others last and last. If you believe, like Dawkins, that miracles are statistical variations, then go ahead and believe that a cow could jump over the moon. Here the blindfolded man doesn't even swing. There is no home run without faith that it can be done.

Alternatively, knowing mankind will have doubts concerning His existence and also responding to mankind's pleas, we can recognize that God has sometimes intervened in the laws He created to aid man.

- Against all odds, can you believe that God so loved mankind that He created for us a home, that, after mankind rejected God by choosing to do things our way, He sent His begotten Son, Jesus Christ, so that all who believe in Him should not perish, but have eternal life, and that Jesus left behind Him a Church, as a pillar and foundation of Truth, to nourish and guide those who do believe?

The State of Our Times

Due to our current state of busyness, and lack of seeking the meaning of life, we have lost many of our fellow adults and much of our youth from the faith, because we, for the most part, have been trained to ignore and to emphasize the science side of life. The questioning side which looks into meaning, such as philosophy and religion, has been placed in the background. Many who were raised on the story of a 6-day creation and the story of the fall of Adam and Eve as a true literalist event now reject religion. Religion cannot be taught as science. Genesis continues to teach great truths about who man is and why he is here, and about the one who created us. The stories are there to give meaning and understanding of the WHY, not the HOW.

Science is being taught in our schools as a religion, that is, a system of beliefs. This is commanded by the government which does not allow the transcendent side of life to be taught. Teaching the transcendent side does not mean teaching religion, particularly a state religion. I certainly agree there can be no state religion. However, many school districts will not even allow the word God be spoken. Since science addresses the world and life as physical things, the subliminal message is sent out that God is not needed. Valedictorians cannot even express their thanks to an entity above and beyond science. Who is it, other than atheists, who believe those listening would be scandalized if they found out the valedictorian believed in a greater power? So, not only must the science be taught, but the limitations of science must also be taught. HOW is as far as science can go. Our students should be allowed to express whatever beliefs they have in appropriate ways.

Atheists do not not want to allow even the slightest opening against materialism. Thus, I can see why atheists insist there is no Body and Soul, but only the material which comprises the Body, for, if one has spiritual and physical components, then atheistic materialism falls apart.

Scripture tells us in Romans 1:20 "*Ever since the creation of the world his invisible nature, namely, his eternal power and deity, has been clearly perceived in the things that have been made.*" We see all the things that human persons do that animals, even the most intelligent animals, cannot. For instance, human persons have empathy, can contemplate the meaning of life, can know what is right and wrong, can choose to do wrong even when we know it is not in our best interests, and can love. The gift of free will is the greatest gift, next to life itself, which we have been given by God. Furthermore, we can know God, if we but seek Him with an open heart and mind.

Once we admit we have Body and Soul, we have gone beyond physicality. If a whole other spiritual existence is present, then the concept of a spiritual, divine person—God—comes to the forefront and makes sense. If, indeed, God made the Universe for man (and for other alien persons if you choose to believe that), then by definition we are creatures and God is the Creator. We then ought to worship Him and must follow His tenets which are established so that we can become "the best version of ourselves," as Matthew Kelly, noted author, would say.[100]

But man wants no restrictions. We often want to do things, as Frank Sinatra sang, "My Way." We often seek power, greed, pleasure, etc., instead of true love, in the false belief that they can bring us happiness. We can ignore the witness of God before our eyes. But we don't have to be that way. We can seek what we have not sought, we can be open to all truths. We can remove that WILLFUL BLINDNESS we all struggle with.

So, as you go about your daily existence, look around, see the world for more than its particles. Give God His due. See the world and others, including that kitchen table I began my book with, in the FULLNESS OF REALITY. Seek and you shall find!

Things to Think About

1. For fun, can you make an estimate as to how likely that a blindfolded man, not knowing when or where a baseball is being thrown, could swing and hit the ball out of the ballpark? Assume he is physically able to and that all pitches are strikes.

---

[100] Kelly, Matthew, *Becoming the Best Version of Yourself,* CD, Dynamic Catholic, at dynamiccatholic.com/becoming-the-best-version-of-yourself-cd-bulk

2. Has your concept of what constitutes reality expanded?

Appendix A
Discussion Questions

Discussion questions are meant for discussions! There is not necessarily any one answer to these questions. Here are some of my thoughts on the questions asked:

Chapter 1

1. Compare the literal versus the literalistic approaches to interpretation. What are a few of the literary genres one finds in the Scriptures?

Response: The first approach to reading Scripture is always to try and determine what the actual words are saying about a situation. The story of Adam and Eve and their fall is necessary to understand the need for a Savior and the story of salvation history. The story line concerning a forbidden fruit is powerful and easily understood by the intended audience. However, the literalistic details, including the words actually spoken by God, Adam, and Eve are not the point. The literal revelation expressed in those conversations are the point. The story of Adam and Eve tells us much about the nature of mankind, how the possession of free will, so necessary in order to be able to love, can also lead us to sin, and how God always remains faithful and loving to us. To help us out to realize the truth behind the story, the existence of an Adam and Eve, and thus the common ancestry of all of us is backed up by science.

So, what are some genres? Here are a few: narrative, poetic, satiric, lyrical, biographical, rhetorical, allegorical, prophetic, parable, apocalyptic, lettered, proverbial. Etc.

2. If the first 11 chapters of Genesis are written in a mythopoetic fashion, how does that affect the meaning of those chapters? Does that mean the events described are not real?

Response: The events are very real, but the specifics of the event are not known. The events (creation, the fall, growth of sin in the world, etc.) are written in a manner which can be understood by the reader. For instance, historically, we know the existence of floods over large areas of land in the past, such as the filling of the Black Sea, could that have been the basis of the whole world's being flooded in the story of Noah. This does not mean that there wasn't a Noah. However, the story of Noah certainly indicates that without God's being part of our culture, the world will progress deeper and deeper into evil, something we see more and more of today. The details of the procurement of every type of animal are not the point of the story and is not even possible considering the existence of so many species on the various continents.

3. Pick any object and construct at least four realities about it. Is a physical description more real than non-physical realities?

Response: Let's take a Basketball and look at various realities.

Referee's reality: A basketball is an air-filled spherical ball of circumference of 22.0" to 29.5". It's made up of various materials. Its inside bladder, which holds the air under pressure, is typically made using butyl rubber. Its outside material can be rubber or leather. The outside is ribbed such that there are 2 lines resembling the equatorial line and a longitudinal line, plus anterior and posterior skew lines which wrap around within the top and bottom halves of the ball approximately 1/8" in width to aid in gripping the ball. Non-standard design or material may not be used in refereed game.

Historical reality: Historians can describe the development of the basketball and of the game. Certain basketballs reflect the times and the stars of the various sporting eras. Certain basketballs have special significance such as ball used to win a championship, or a red/white/blue basketball used by the American Basketball Association professional basketball league.

Sentimental Reality: A basketball can have a special significance because the ball was given to a kid by his father when he was a little boy. Even holding the ball brings back warm images of his father. No other ball has that reality.

Artist's Reality: The choice of colors invokes different thoughts in those who behold the ball. The spinning motion of the ball presents an artistic expression as it traverses through the air, or rebounds in a certain sound from the floor.

4. Can one determine the full meaning in anything if one restricts his or her understanding to materialism?

Response: Materialism is simply too limited to convey anything more than a partial meaning. The most important things in life are non-material. Materialistic reality is cold, superficial, does not reach into the heart. Materialistic reality is important in order to discover how things work and how to use things to our best advantage. *How* is an important question. *Why* is an even more important question. We can live without knowing how anything works, but we live a very limited life if we don't know love, peace, justice, etc.

Chapter 2

1. How would you explain, if you had the job of explaining to an audience who knows nothing of science nor of the origin of the Universe, the Earth, and the Earth's plants and animals? Your explanation must explain not only the how, but also the why.

Response: One gains knowledge, not by suddenly understanding a new idea, but rather by building on what one already knows. The Scriptures were written as revelation of God, revelation of the nature of man, and revelation of the relationship between God and man. Part of that revelation is a description of how man and the world around him began. To get the meaning behind those events, one could not describe just the physics of the events. Instead, the inspired human authors chose to use visualizations of realities that man at that time could understand. This includes the description of stars, seas, land, plants, animals, etc. It is far more important to describe the effects of feeling the warm sun on one's face, than how the sun generates heat. So, too, it is far more important to know that the heavens were created in an orderly way, then to know exactly what the science of that event entailed.

2. What would be the purpose of describing science to such an audience? What is the real purpose that the Scriptures were written anyway?

Response: The Scripture writers also recognized that man is different from the animals and possessed a transcendent nature. They wanted to show this transcendent nature which allows one to love, to trust, to see beauty, to seek truth. In addition, they visualized men and women as being different from each other yet equal in nature to each other. They knew that mankind's nature has a weakness which leads one to prideful, sinful actions. Thus, they described in Genesis stories which showed the misuse of free will and its results. They used what they knew to be the nature of mankind to account for how man's fall came about. Science does not explain the transcendent nature of man.

3. Many Evangelicals/Fundamentalists get around the six 24-hour day quandary by acknowledging that the word for day in the Hebrew scriptures "yôm" can also translated epoch or eon. Thus, they can maintain a literal inerrancy of Bible by saying, "Well, it could

have been a long time." How does this approach still misunderstand the Inerrancy of Scripture? How does it miss the point of Scripture?

Response: Even if one applies the concept of an epoch in place of the 24-hour day, the story of creation does not match the story of creation as seen in the evidence from the remnants of the Big Bang. For instance, on Day 3 in Genesis, God creates vegetation but doesn't create the sun and the moon until Day 4. Even if one allowed an epoch to be a variable long time, say a billion years, this would not make the literalist interpretation be true. We know that vegetation developed through a long period and certainly needed the light of the sun for photosynthesis. Again, the purpose of Scripture is to explain the Why, not the How. The Bible is inerrant means what was written is the truth intended by the authors to help us understand and develop a relationship with God, resulting in our salvation.

Chapter 3

1. Can the development of the Universe be both random and orderly?

Response: Are random and orderly oxymorons? In classical physics, the interactions of materials were thought to be predictable. In fact, during the so-called Enlightenment, there were those physicists who believed everything was deterministic. One could, if all the parameters were known, predict both the past and the future of anything, but quantum physics has taught us otherwise. All interactions in the science world are based on probability. We can project how far a cannonball can travel through the air, not because that distanced is predetermined by the laws of physics, but because the laws of physics teach us that the probability of the cannon ball doing anything other than what we calculate is so minute that one will never see it happen. Similarly, we know that all the air molecules in a room COULD statistically move to one side of the room, yet it never happens. However, the interactions of the air molecules and the cannon ball, while being probabilistic, are not random. Individual interactions are not 100% predictable, but all follow the laws of physics which allow for probability that a certain event will have a certain outcome. Therefore, they are ordered to the laws of physics. Random events would follow no rules at all and would occur for no reason. Thus, the Universe IS NOT random and orderly. It operates probabilistically according to the defined laws of physics—laws created with an intent in mind, laws which give us order.

2. In ancient Greece, two philosophers, Aristotle and Zeno, had a disagreement about whether motion was continuous or discrete (that is in steps). Zeno said discrete, Aristotle said continuous. Basically, in my version of the argument, Zeno would have said that if a runner moved 4 feet in one step, then half that distance (2') in the next, then half that distance (1') in the next, etc., he would never cross the finish line. Aristotle would have said motion must be continuous as the runner obviously does cross the finish line, so Zeno you are one crazy philosopher! Aristotle won the minds at that time, but he was wrong. How does there being a minimum distance interval show that Zeno was right?

Response: If Aristotle had understood quantum physics, he would have understood. When the runner's steps which are constantly cut in half, reach the very minimum distance which quantum physics allows, the choice is binary—either the runner moves 0.0 distance or the runner is forced to continually move forward at least with the minimum distance each time, and, thus, the runner eventually leaps across the finish line with a step of the minimum distance quantum physics allows. Thus, we see the philosophical truth is fully reconciled when the truth of science adds-in information to the argument. Initially, perhaps one would have thought, science and philosophy don't mix, but as we find out, there is but one truth and irreconcilable differences must somehow be reconciled if, indeed, they are both true.

3. There are so many requirements that force strengths, charge ratios, mass size, etc., must be basically exactly as they are if man is to exist. Some say that we were just lucky; others say we weren't just lucky, but that this happened, therefore, it must be achievable. So why worry about it? Why don't these answer the real question?

Response: Saying we were just lucky is hardly scientific. We understand that the Universe is orderly, there, it must be understandable. Yet, we come to a point where science tells us that it is unreasonable for all the steps to continue to occur because the probabilities are just too small considering the young age of the Universe. We are left to understand that there must be a super-intellect, as Fred Hoyle said, outside the Universe, who has "monkeyed around with the physics." As we have seen in discussing different realities, no one has shown that science is the only reality, or even the only controlling reality. Logic says there must be more.

Chapter 4

1. What new knowledge have you learned about the heavens and the Earth since you were a child? How much was this knowledge gained by your figuring it out?

Response: As a child, I learned that there were nine planets rotating about the Sun—Mercury, Venus, Earth, Mars, Jupiter, Saturn, Uranus, Neptune, and Pluto. Now, we have been told that Pluto is too small and its orbit is too offbeat, and it has been diminished to being called a dwarf planet. There is a Planet X which hasn't been found yet, which is 10 times the size of the Earth and 300 to 1000 times farther out than the Earth's orbit. Hmmm! Also, we now know about binary stars, quasars, black holes (fuzzy or not), curved space, an incredible number of galaxies, etc. How much of this did I figure this out? Zero percent. In reality, our knowledge is nearly all dependent on the trust of others.

2. The chapter noted some special properties of water. Can you name any others?

This is just another example which shows how special our Universe and its laws are. Here are some unique and necessary properties of water that are just right:

Adhesion — Water sticks to things.

Buoyancy — Water tends to exert an upward force on objects that are immersed.

Capillary action — Water can move through narrow porous spaces.

Cohesion — Water sticks to itself due to the POLARITY of charges within water.

Surface Tension — Water exerts a force on the particles on its surface.

Solvency — Water is a solvent. All kinds of solid chemicals dissolve in water. Without this property, how would important molecules get in and out of our cells? Acids exist in a water base. Need that to dissolve foods. Our body is mostly water!

Reflectance — In the sky, clouds reflect light away from the earth, keeping the Earth cooler, yet insulates the surface of the Earth, keeping the Earth warmer.

And to think, that water can be solid, liquid, or gas in the narrow temperature range we have. Incredible!

3. Some people just seem to be lucky. Is being lucky really a personal attribute?

Response: Let's face it, being lucky is not too reliable of a personal trait. It's not that one can rely on your friend to be lucky when he goes to the casino in order to win money at the slot machines, and, thereby, pay off his debt to you. I, personally, would rather rely on the trustworthiness of my friend to pay me back. On the other hand, my grandfather used to win 50/50 raffles a couple of times a year. I'm still waiting for my first win. Of course, I rarely enter them. I wonder, is there a correlation?

When you look at the relative odds of the various one-time events which all had to happen at the right place and time as part of cosmic evolution, and even scientists claim we are just so lucky, one has to wonder, "Can luck really explain it?" At what point does one finally concede that no way is any event this lucky. A number I have heard is 1 in 10^{50}, or 1 in 100 trillion trillion trillion trillion. The numbers associated with cosmic evolution are way beyond this, yet some atheists still have faith that random luck must be the one and only answer.

However, some of us are lucky. After all, I certainly was lucky to be born to the parents that I had. I can't say I had any choice or merit which allowed me to pick my parents. And what of those people we call lucky? Even if someone is said to live a lucky or charmed life, there is the implication that someone is applying the luck or charming upon them. Somehow, one just cannot get around there being an *other* who is messing with your life. Maybe my having the parents that I had was not just a lucky event, but part of an eternal plan for me!

Chapter 5

1. Why is the Origin of Life not an evolutionary process?

Response: Evolution is the development of something already present. Thus, an idea can involve, a physical process can develop, life can change. However, when it comes to the initiation of life from non-life to life, there is no cumulative improvement. With the Origin of Life there is needed more than a lightning bolt happening to strike a bunch of chemicals that, for no reason, have arranged themselves in just right alignment. The simplest of conceivable life is utterly complex. Processes have to already be in place. It is very much like the Origin of the Universe. Cosmic evolution could not have started just because suddenly there was a blast of energy. Cosmic evolution required the presence of the laws of physics and its corresponding forces (which are not physical) to exist at the instant of the "Big Bang". Similarly, the Origin of Life requires the presence of information (which is not physical) at that moment of initiation.

2. The concept of the requirement for information for the existence of life is a game changer. Why is that?

Response: When one speaks of life, one is not just speaking of molecules, somehow, assembling and replicating themselves through the laws of physics. There must be, as noted in response to question 1, a non-physical element present—that of information. Materialists can no longer remain as materialists. Materialists must include the non-physical information. They must become Materialists Plus.

3. How are computers conceptually similar/different from DNA?

Response: Both computers and DNA have a language associated with them. As discussed in the chapter, DNA makes use of four possible characters and 20 possible words. The characters are grouped into sets of three and are called a codon. Computers makes use of two characters. The characters, called bits, represent 1 or 0 (on or off). The bits are then grouped into sets of eight and are called a byte. An 8-bit byte can represent 256 possible words. Normally two bytes are used to represent a command.

DNA uses codons aligned in a certain order to program the functionality of the cell. Computers use bytes arranged in a certain order to program the functionality of the computer. Humans are used to program the computer. Materialists believe that randomness programmed the first cell, then future cells and all life evolved from that first cell. Fortunately, so far, computers have not been able to create other computers without the help of humans to program them!

Chapter 6

1. Why do you think the time factor concern has only been addressed in the last 50 years or so?

Response: For one thing, six-day Creationism was part of the belief system of a larger percentage of the population. However, the discovery of DNA as the primary instrument of genetics led one to understand and formulate the language of DNA. At one point, some researchers believed that the very design of the DNA structure and the corresponding design of amino nucleotide bases mandated that the nucleotide bases would naturally align properly. With the understanding of the language of DNA, the exact order of the nucleotide bases became a necessity. Since there would need to be a random, cumulative process, the question became how long would it take to make just the right sequence. This was followed by the thought as to how long that process would reasonably take.

2. What are your thoughts on Panspermia?

Response: The concept of Panspermia does add more time on the clock, as compared to the 600 million years from Earth formation to known life. But Panspermia never addresses the real question of how life emerges from inorganic materials. Besides, the age of the Universe is still too young to allow life to develop even if the process started on Day 1. I guess one would have assumed that life jumped from one older Universe as part of the Multiverse.

3. When asked how life began, Richard Dawkins indicated that no one has any idea. Is that true?

Response: Once again we see a fixation on materialism as the only source of Reality. Man has long recognized the reality of a God. All cultures have the concept of a transcendent being. The concepts may be off-base for some of the religions of the world, yet they DO have an idea how all this occurred.

Chapter 7

1. We hear about physical disasters such as earthquakes, volcanic activity, hurricanes, ice ages, global warming, etc. Has looking at past disasters changed your thoughts about the development of life?

Response: There are those who believe everything happens for a reason. Certainly, looking at the Earth's past seems to reinforce that thought. Not only did the past events occur but the timing of the past events seems to be at the right time for the development of a full ecological system we have now.

2. What are some evidences you have noticed that support Biologic Evolution?

Response: Evolution is not a modern invention. The ancient Greeks and others noticed the similarities between various animals and suggested they had common ancestors. Less intuitive is the thought that plants, fish, reptiles, amphibians, mammals, plus unicellular bacteria, viruses, etc., all could come from a common line. Yet, evidence exists such as the common sequences of DNA discussed in the chapter that either there was a common path or, perhaps, that is the only way life could exist. The biggest argument seems to be whether there were necessary interventions such as the time of the Cambrian Explosion. However, no matter how all the different forms developed, I am glad they did.

3. Because some biologic developmental steps seem to follow evolution, does that necessarily mean all development is evolution?

Response: When solutions are not definitively proven, it is best to keep an open mind and examine all evidence and solutions; otherwise, you might end up with a partial view of Reality. God could have chosen evolution to be the sole process or not. Science is not able to say one way or another, but just theorize, and show the evidence that is seen.

Chapter 8

1. Could a robot ever be a person?

Response: There are individuals who believe that someday robots will become rational persons with free will. Richard Dawkins is one of them. Personally, I recognize that a nation of robots could be programmed to do all the steps necessary to mine, design, and manufacture other robots. In effect, they would be reproducing themselves and doing all the things necessary to be called living. After all, there is no reason that all life forms must be carbon based. Our planet certainly seems to be all carbon based, but others insist silicon might also work. However, that doesn't make one a person. A person needs to also have free will. Robots are all programmed. Their decisions are based on the programmer logic that was put into the program. Because a robot can beat me in a game of chess does not make the robot a person; rather, it is a finely programmed robot developed by a human's insight that beat me.

2. Could Near Death Experiences just be a chemical manifestation originating in the brain?

Response: Well, for certain cases, what an individual believes is a real NDE could in fact be a biochemical effect. However, the individual may not recognize the difference between an NDE experience and simply seeing Uncle Harry in a dream. However, some medical scientists try to put all those who claim to have an NDE in that box. There are differences. For instance, a common difference between drug-induced memories and NDE events is that NDE events are quite lucid compared to events that happen under the influence of anesthesia or drugs. Of course, there may be other explanations available. Humanity being what it is, in some cases, a person for their own personal reasons may fabricate a story. But just because a few cases are explained away doesn't allow one to throw out all the thousands of other cases that have been reported. On the other hand, persons undergoing an NDE event often report knowledge of happenings which they have no reasonable way of knowing. For instance, how could a drug-induced state allow one to see and hear things that happen at a place where their body is not?

3. Our "vision" and "hearing" senses are limited by the earthly vessels we call bodies. What of our other senses? How might being outside the body enhance our other senses.

Response: All of our senses are limited by our body capabilities to detect. NDE'rs often speak of hearing sounds and seeing colors they never knew before. One would think that this could apply to other senses also. Certainly, our ability to smell is much more reduced than many animals as is our sense of touch. I have not investigated this, but I believe unusual odors or touch sensations must happen to those who go outside the body.

Chapter 9

1. Are the 10 Commandments a reflection of man's conscience?

Response: The 10 commandments are laws written into the heart of man. All people can come to the realization that parents are to be honored, killing another is wrong except in self-defense, one should be faithful to one's spouse, it is wrong to steal, it is wrong to tell lies, it is wrong to covet your neighbor's wife or your neighbor's goods. Once one realizes there is a God then one must worship and obey Him. This is natural law which is embedded into man's conscience—at least initially.

2. If the conscience is a genetic biological code, would it not be subject to mutation?

Response: If man is simply an arrangement of atoms, then if that arrangement would get changed, so, too, would the conscience from generation to generation, but it hasn't.

3. What about love? Is it genetically controlled?

Response: Love is more than emotions. Free will, which cannot be controlled by genetics, is required to be able to love. Therefore, love is independent of the body and is not genetically controlled.

Response: True love is always a choice. Physical love is not love but a fruit of true love between persons. How much physical desire is present may be a function of biology, but love is beyond the physical. By definition, if love were genetically controlled, it would not be love. To love, one must possess free will.

Chapter 10

1. What are for you the most intriguing aspects of the Eucharistic miracle at Lanciano?

Response: There are many things to consider—why the host has not decayed, why the blood is that of a Middle Eastern man, etc., but my favorite is how could one attach muscle to a rim of bread. The interface is smooth with no evidence that the muscle is a separate piece from the bread. It's just not possible.

2. Regardless of whether the Shroud is the burial cloth of Jesus, what are for you the most intriguing aspects of the Shroud of Turin?

Response: Again, there are many things to consider. There is no body image under the blood marks. The density of the coloring of the image contains 3-dimensional information. The image only makes sense if the image was obtained when the shroud was wrapped around a body, the image shows signs of rigor mortis, components of the blood known only to modern man are present, etc., but my favorite is the lack of any known modality which can produce such an image. The presence of an image is miraculous enough, much less an image which depicts many of the known characteristics of the torture and death of Jesus Christ.

Chapter 11

1. Can you imagine walking 13 miles one way to receive catechetical instruction? What extraordinary efforts have you made in your life to learn more about your faith?

Response: I'll let you reflect on this for yourself. I know I have reflected on this and found myself wanting!

2. What physical aspects of the Tilma of St. Juan Diego are so remarkable and unexplainable?

Response: As with the Eucharistic miracle at Lanciano and the Shroud of Turin, there are many physical characteristics of the tilma of Juan Diego which are simply unexplainable. The tilma became a miraculous cloak somehow. Its longevity, its resilience, its history, its inherent codes and images, etc., are quite remarkable. However, from a physical perspective it is the Polaroid-like nature of the image without any Polaroid chemicals that is so miraculous.

3. What are the spiritual aspects of the Tilma of St. Juan Diego that are so remarkable and unexplainable?

Response: We all know the story of the United States and the fate of the Native Americans. While some attempts were made to purchase the lands for ridiculously low prices, for the most part the land and the people were just subjugated and put on reservations. Not so in central America. The racism seen in the United States between the colonizers and the Native Americans and the slaves brought from Africa did not occur to any great extent in Mexico. Nine million Aztecs and other Native Americans converted to the Catholic faith. The conqueror and the conquered became integrated, intermarried, and merged together.

Chapter 12

1. Can you imagine the anguish the three young children of Fatima must have endured? They must have been under tremendous pressure. As a young child, could you have withstood onslaught of the negative attention, they must have felt?

Response: One of the major arguments for the reality of the Resurrection of Jesus of Nazareth is that the Apostles would not have willingly suffered extreme torture and death for a lie. Someone would have folded and said that the story was made up. The same is true for the children of Fatima. It is hardly imaginable that all three children would have been able to maintain a lie under such pressure. It is far more likely that the children would have lied and denied the apparition in order to escape the pressure.

2. *The Miracle of the Sun* was seen within a 20-to-25-mile radius of the Cova de Iria near Fatima. The rest of the world did not see such an event. Is it possible that such a thing physically happened to the Sun?

Response: Every year at Christmas, there is a claim that the Star of Bethlehem which led the astrologers to the Holy Family—Jesus, Mary, and Joseph—was a real celestial event, usually involving the conjunction of planets, such as Jupiter and Saturn, thereby, making a brighter than normal light in the night sky. How this would lead one to a particular house in Bethlehem is beyond understanding. However, it is recorded that such a light in the sky did just that. The event at Fatima was witnessed by over 70,000 people. To deny that is to ignore a very large data set of evidence. From a physics point of view, under the laws of physics, the Sun could not have danced in the sky and not have had other disastrous effects. However, God in his almighty power, could cause the 70,000 (and the 3 wise men) to see what he wanted them to see in their mind. Mass hysteria could not be the answer. I accept the vision that they saw plus the documentation of the other miracles on that day as testified by 70,000 people that it happened and acknowledge that it is a mystery.

3. Many of the bodies of saints have decayed somewhat since they have been discovered and put on display. Some have been restored through the use of wax and other means. Does this mean they really aren't incorruptible? Does this mean there has been no paranormal event going on?

Response: It is acknowledged that there are some rare occasions in dry, arid locations where bodies have not corrupted. Those necessary conditions do not fit the environment found at Lourdes and other places. The bodies should have decayed. The fact they didn't is the issue. The question as asked twists around the real point, which is, were the bodies incorrupt when found?

Chapter 13

1. For fun, can you make an estimate as to how likely that a blindfolded man, not knowing when or where a baseball is being thrown, could swing and hit the ball out of the ballpark? Assume he is physically able to and all pitches ae strikes.

Response: Let's assume the strike zone is 4 ft high, the homerun hitting part of the bat covers a fourth of the plate (in 4 swings he would cover the plate in to out), the batter is able to swing once within a 3-second time frame without getting tired. Furthermore, the sweet part of the bat is only 1/4" of the barrel of the bat. The pitcher throws only strikes over the plate over 4' area height. Each pitch is made somewhere within a 3-second window by our tireless pitcher, so every 3" a ball will be at the perfect spot to be hit. There is only 0.1 second out of every swing where the bat is moving with the right speed and the barrel would be at the right position extended out over the plate.

The likelihood of hitting the sweet spot at the right time becomes:

1/4 (bat correct inside to out) x 0.25"/48" (correct height) x 0.1" (perfect timing)/3" between pitches x 1 swing/3 seconds = ~1 per 69120 swings. So, about every 70,000 (7×10^4) swings our tireless batter would be able to hit one out of the park! Now have him hit 2 homeruns. The odds become 1 in every 4.8 trillion swings. The time for 2 homeruns would be 4.8 trillion swings * 3 seconds per swing = 14.3 trillion seconds, or about 23,000 years. If this seems improbable, then imagine working with probabilities of 1 in 10^{50} or greater!

2. Has your concept of what constitutes reality expanded?

Response: You'll have to answer this for yourselves.

Appendix B

List of Figures

Figure	Description/Permissions	Page
2.1	G. K. Chesterton Figure 2.1 G.K. Chesterton, early 20th century English writer, philosopher, lay theologian, and literary and art critic.	21
2.2	A drawing of an atom.	39
3.1(a)	A missile explosion.	51
3.1(b)	A fireworks explosion.	51
4.1	Irregular, Elliptical, and Spiral Galaxies	62
5.1	Information contained by placing letters onto a string.	86
5.2	Schematic of a single-stranded RNA molecule.	88
5.3	Basic protein structures.	90
5.4	Beads on a string giving a command.	91
7.1	Catastrophic events which have occurred since 542 million years ago.	109
7.2	Ediacaran fossil.	113
7.3	The human and chimpanzee chromosomes.	125

9.1	The Conscience By François Chifflart (1825-1901).	167
10.1	Monstrance containing Eucharist miracle of Lanciano, Italy.	179
10.2	Enlargement of the image of the Lanciano host/flesh.	179
10.3	The Shroud of Turin on Display in 1931, in Turin.	185
10.4	Striated straw-colored fibers across the linen threads.	188
10.5	Posterior view of the scourge marks across the back of the man in the Shroud of Turin.	191
10.6	Typical Roman flagrum used for scourging.	192
11.1	Image of Our Lady of Guadalupe.	201
11.2	Map of area near Tepeyac Hill.	203
11.3	Typical Aztec-style of painting.	206
11.4	The image of Our Lady of Guadalupe, the painting of Our Lady of Mercy, a later painted copy of Our Lady of Guadalupe, and a sculpture of Our Lady at the Monastery of Guadalupe of Caceres.	208
11.5	The *Nican Mopohua* and an image contained wthin the death certificate of Juan Diego.	211
11.6	A modern-day Agave plant.	213

11.7	Photo of the face of a young man.	216
11.8	Images seen within the left eye of Our Lady of Guadalupe, as they are blown-up and computer enhanced.	217
11.9	Further Processing of both eyes, showing slight shifts in the position of the figures due to separation of the eyes.	217
11.10	Further processing of the image of the left eye with the family image removed, leaving behind a shadow of the Our Lady of Guadalupe.	218
11.11	The metal cross which stood next to the image of Our Lady of Guadalupe as it looked after the explosion in 1921.	220
11.12	The Image of Our Lady of Guadalupe overlaid over a map of the volcanoes of Mexico.	220
12.1	Jacinta Marto, age 5, Francisco Marto, age 7, and Lucia dos Santos, age 8 (in 1915).	228
12.2	Images of the crowd at Fatima before and during the Apparition Oct. 13, 1917.	233
12.3	Bernadette Soubirous in 1858.	245

Appendix C
Suggested Reading

Books

Barr, Stephen M. *Modern Physics and Ancient Faith.* University of Notre Dame Press, 2013.

Davies, Paul C. W. *The Accidental Universe.* Cambridge: Cambridge University Press, 1990.

Hawking, Steven, *A Brief History of Time.* New York: Bantam Books, 1988.

Hawking, Stephen, and Leonard Mlodinow. *The Grand Design*, New York: Bantam Books, 2010.

Lewis, Geraint, and Luke Barnes. *A Fortunate Universe: Life in a Finely Tuned Cosmos.* Cambridge: Cambridge University Press, 2016, 3rd Printing, 2018.

Ross, H. *Improbable planet: How Earth became humanity's home.* Grand Rapids, MI: Baker Books, 2017.

Polkinghorne, J. C. *Quantum Physics and Theology: An Unexpected Kinship*. SPCK, 2007.

Ross, H. *The creator and the cosmos: How the latest scientific discoveries reveal God.* Covina, CA: RTB Press, 2018.

Spitzer, Robert J. *New Proofs for the Existence of God: Contributions of Contemporary Physics and Philosophy.* Grand Rapids, MI: William B. Eerdmans, 2010.

Axe, Douglas. *UNDENIABLE: How Biology Confirms Our Intuition That Life Is Designed.* HarperCollins, 2017

Moody, Raymond A. *Life after Life.* New York, NY: HarperOne, an Imprint of HarperCollins Publishers, 2001.

Collins, Francis S. *The language of God: a scientist presents evidence for belief.* Free Press, 2007.

Coyne, J. A. *Faith versus fact: Why science and religion are incompatible.* NY, NY: Penguin Books, 2016.

Dawkins, R. *Climbing mount improbable.* New York, NY: W W Norton, 2016.

Dawkins, R. *The blind watchmaker: Why the evidence of evolution reveals a universe without design.* London, UK: Penguin Books, 2016.

Dawkins, R. *The God delusion.* London, UK: Black Swan, 2016.

Long, Jeffrey, and Paul Perry. *God and the Afterlife: The Groundbreaking new evidence for God and near-Death Experience.* HarperOne, 2016.

Antonacci, M. *The resurrection of the Shroud: New scientific, medical, and archaeological evidence.* New York: M. Evans, 2001.

Franciscan Friars of the Immaculate, eds. *A Handbook on Guadalupe.* New Bedford, MA: Ignatius Press, 2001.

Badde, Paul. *María Of Guadalupe: Shaper of History, Shaper of Hearts*. Translated by Carol Cowgill, Ignatius Press, 2009.

Callahan, P. S. *The tilma under infra-red radiation: An infrared and artistic analysis of the image of the Virgin Mary in the Basilica of Guadalupe*. Washington, D.C: Center for Applied Research in the Apostolate, 1981.

Górny, G., Rosikoń, J., & Kacsprzak, S. *Guadalupe mysteries: Deciphering the code*. San Francisco: Ignatius Press, 2016.

Johnston, Francis W. *The Wonder of Guadalupe: The Origin and Cult of the Miraculous Image of the Blessed Virgin in Mexico*. Tan Books, 2011.

Cruz, J. C. *The incorruptibles: A study of the incorruption of the bodies of various Catholic saints and beati*. Charlotte, NC: TAN Books, an imprint of Saint Benedict Press, LLC, 2012.

Haffert, J. M. *Meet the Witnesses of the Miracle of the Sun*. Spring Groveublin, PA: The Society for Christian Civilization,2006.

Zaki, P. *Before our eyes: The Virgin Mary, Zeitun, Egypt, 1968 & 1969*. Goleta, CA: Queenship Pub, 2002.

Websites of Interest

Near Death Experience Research Foundation, online at www.nderf.org.

International Association for *Near Death Studies,* online at iands.org/home.html.

The Animated Genome, online at unlockinglifescode.org/ media/ animations/659#660.

Evolution News, at evolutionnews.org.

Luskin, Casey. "The Fine-Tuning of the Universe." at Evolution News, at evolutionnews.org/2017/11/ids-top-six-the-fine-tuning-of-the-universe.

Index

Abortion 133-134
Adenine.......................... 87
Angel.. 156, 209, 221, 226, 228
Antonacci, Mark......... 188
Asteroid 30, 68, 70, 71, 83, 102, 104, 115-117
Astrologers 237, 302
Astrophysics 140
Atheism...... 10, 12, 57, 249
Atheist.....ii, 11-12, 21-22, 25, 30, 38, 43, 50, 52, 57, 60, 72, 76, 133, 137, 184, 236-237, 240, 248, 262-263, 272-273, 289
Atmosphere. 111-113, 116
Augustine, Saint...... 16-17
Axe, Douglas 120-121
Bacteria 104, 111-113, 294
Bible..... 9, 15-6, 18, 23, 26, 42, 87, 283
Big Bang 31, 33, 35, 44-45, 52, 56, 260, 283, 291
Birds.......... 37-38, 117, 204
Blind i, ii, 2, 10, 14-15, 30, 57, 75, 78, 138, 149-156, 162, 169, 227, 238, 255, 265, 274
Bosons 39
Calcium 114, 244, 248
Cambrian Explosion. 113-115, 118, 269, 295
Carbon.... 57, 66, 67, 79, 111-112, 195-196, 198, 248
Cardiac....... 146-147, 180-183, 199
carnivores..................... 116
Catholicism......... 182, 223
Chemistry... ii, 10, 57, 82, 114, 142, 181, 198, 214
Chimpanzee......... 124-126

Civilization ...71, 104, 156, 268
Clay.........................83, 112
Clinton, Bill...................87
Codon 88-92, 97, 120, 291-291
Comatose.............145, 147
Comte..................248-249
Conquistadors.............202
Conscience .167-171, 173, 175-176, 298
Consecration.......178, 231
Cosmos....................13, 53
Cova ... 228-231, 234, 236-237, 251, 302
Creation.15-16, 19, 23-26, 28, 31, 33-36, 38, 41, 44, 48, 53, 56, 59, 64, 68, 77, 123, 198, 261-262, 269-270, 272-273, 278, 283
Crevices........................103
Crick, Francis.............264
Cystine..........................87
Dark Energy..................46
Dark Matter..................46
Darwin, Charles..96, 108-109, 132-133, 172-173
Dawkins, Richard.......10, 72, 96, 100-101, 105, 108, 115, 118, 120-122, 133-134, 171-173, 238-239, 251-261, 264, 270-271, 293, 295
Death 132, 134, 136-138, 141-144, 149-150, 152, 156-160, 167, 191, 210-211, 240, 242, 296, 300, 302
Dentures.......................146
Design....5, 11, 57, 59-60, 172, 208, 221, 255, 263-264, 269, 279, 292, 296
Determinism 11-12
Devil 202-203, 230, 236
Diego, Juan.202-207, 209-212, 217, 224, 301
Dimensions... 14, 190, 266
DNA.....67, 84-85, 87-92, 97-99, 118, 122-124, 127, 173, 231, 264, 291-293, 295

Ediacaran.....................113
Einstein, Albert... 1-4, 255
Electromagnetism...45, 47
Elliptical.........................62
Embryo 30, 92, 133
Entropy......43, 50-53, 260
Epicurus.......................136
Essence.............6, 178, 269
Eucharist....178-180, 183, 199-201, 224, 250, 259-263, 299-301
Evolution.10, 53, 61, 72, 79-80, 95-97, 100-102, 106-109, 115, 117-118, 120-123, 127, 171, 173-175, 181, 260, 269, 289-290, 294-295
ex nihilo 24, 31
faith....iv, 2, 6, 9-11, 16-19, 22, 40-41, 58, 76, 104-105, 170, 179-80, 182, 184, 207, 239, 254, 265-266, 270-272, 278, 289, 298
Fenwick, Peter.............142
Fetus.....................133-134
Feynman, Richard 5
Force...8, 45-49, 57, 59-60, 68, 93, 128, 255, 260, 267, 285-286, 288, 290
Gaskiers Glaciation.....112
Gene...... 92, 167, 171-173, 175
Genesis..15-16, 19, 24-26, 34-37, 39-40, 43, 75, 77-78, 123, 203, 272, 278, 282-283
Genome..........10, 123-124
Godzillion.. 32, 53, 55, 98, 255, 260, 267, 269
Gravity.45, 48-49, 70, 139, 263
Grunberg, Peter.............10
Guanine..........................87
Hallucination.63, 161-163
Hamlet.........................119
Hawking, Stephen.. 10-12, 38, 52, 255, 263
Heaven...17, 23, 34, 38-39, 59, 132, 145, 157, 204, 219, 230, 265, 282, 287
Hemisphere 65
Herbivores 116

Homer.................. 135-136
HOW...34, 40-41, 78, 262, 263-265, 272
Hoyle, Fred.......31, 57-59, 286
Hubble, Edwin 31
Illusion 2, 3, 162
Incorrupt...240-244, 250, 303-304
Incredulity....96, 104, 122, 138, 238
Isotope 27-28, 79, 102
Jesus....17, 27, 32, 78, 139, 157, 160, 178-180, 182, 183, 191-199, 202, 221, 226-230, 232, 234, 237, 240, 271, 299-300, 302
Kobilka, Brian Kent...... 10
Lemaître, Fr. Georges... 31
Lanciano....178-180, 182-184
Late Heavy Bombardment .71, 80, 102
Lava 28, 115, 241
Lewis and Barnes ... 44, 72, 128
Literal 16-19, 44, 277, 283
Literalist. 16-19, 26, 34, 272, 277, 283
Luck...59-61, 72, 98, 100-102, 128, 207, 286, 289-290
Matter.....1, 12, 21, 44, 46, 49-50, 79
Moon.....17, 29-30, 38, 45, 65, 67, 69-71, 102, 209, 234, 257, 259, 271, 283
Multiverse...25, 31-32, 76, 78, 127, 267, 293
Murray, Joseph.............. 10
Mutation......97, 108-109, 118-121, 132, 261, 298
NDE......137-138, 140-144
Neo-Darwinism (-ian).96, 108, 132
Neoproterozoic .. 110, 112
New Testament 87, 170, 191
Nobel Prize.5, 10, 214, 264
Nucleotide. 87-88, 98-100, 292-293
Old Testament............ 170
Orbit ...5, 39, 56, 58, 64-65

Our Lady of Fatima .. 227-228, 231, 234, 236-238, 250, 302-303
Our Lady of Guadalupe 201-202, 206-213
Oxygenation 110, 112
Ozone 67, 82
Panspermia 103, 293
Paramagnetic 29
Phaedo 136
Photosynthetic....67, 112-113
Radiation...63, 67, 71, 82-83, 117, 197-198
Radioactive..27-28, 48, 62, 68, 71
Radiocarbon 196
Reptiles........................ 294
Resurrection .17, 198, 240, 302
Resuscitation 147
Revelation...15, 239, 254, 266-267, 277, 281
RNA...80, 82, 84-85, 87-92, 95, 97-100, 102, 120
Sagan, Carl.............. 13-14
Saints ... 244-245, 250, 303
Saturn 287, 303
Schönborn, Cardinal Christoph 17-18
Scripture.15-19, 23-24, 26, 34, 40, 43, 75, 87, 170, 226, 267, 273, 277, 281-283
self-replicating............. 105
Shakespeare 118
Shroud of Turin . 184-185, 191, 196, 198-199, 201, 213, 299, 301
six-day 24, 26, 292
Soubirous, Bernadette 243, 245
Spontaneous 197
strong force............. 47-49
Sun7, 11, 17, 36, 38, 45, 58, 64-67, 69, 70-71, 82, 112, 117, 209, 219, 234, 237-239, 268, 281, 283, 287, 302-303
Tectonic 71, 111, 113
Temperature....56, 65, 67, 69, 71, 81, 102, 111-112, 116-117, 212, 288

Theology......................261
Tidal..............................67
Tides..............................70
Tilma...202, 206-207, 209, 212-216, 218-219, 221
Universe.....262-263, 266-269, 273, 281, 283-284, 286-287, 290, 293
Uranus..........................287
UV.........................67, 83
Vatican II....................168
Virgin Mary......204, 209, 227, 245, 256
WEASEL......119-120, 261
WHAT............34, 40, 254
WHO.......34, 40, 254, 262
WHY......34, 38, 40, 78, 254, 262, 265, 272
Willful Blindness...10, 14-15, 138, 227, 238, 255, 265, 274
Yucatan......................116

Made in the USA
Monee, IL
17 July 2021